W9-AZX-507

Photo by Tsuyoshi Nishiinoue and Orion Press
Raging Forces: Earth in Upheaval
This 30th-Anniversary Edition belongs to...
Name
Date

Charles Caratini/SYGMA

In February 1995 a rescue worker uses a still operating public telephone in Warcq, France.

RAGING

FORCES

Earth in Upheaval

Prepared by
The Book Division
National Geographic Society
Washington, D.C.

PRECEDING PAGES: Lava cascades down the slopes of Sicily's Mount Etna in 1993. More than 200 recorded eruptions make Etna one of Europe's most active volcanoes.

ABOVE: Crashing through patio doors, water pours into a living room as giant waves pound a beachfront home on Jupiter Island, Florida, in October 1991.

David Lane/*PALM BEACH POST*; Roger Ressmeyer/CORBIS (preceding pages)
W. Balzer/WEATHERSTOCK (following pages)

RAGING FORCES Earth in Upheaval

Contributing Authors

Leslie Allen
Thomas Y. Canby
Ron Fisher
Noel Grove
Tom Melham

Published by
The National Geographic Society

Gilbert M. Grosvenor
President and Chairman of the Board
Michela A. English
Senior Vice President

Prepared by
The Book Division

William R. Gray
Vice President and Director
Charles Kogod, *Assistant Director*
Barbara A. Payne, *Editorial Director*

Staff for this book

John G. Agnone
Project Editor and Illustrations Editor
Martha C. Christian
Text Editor
Cinda Rose, *Art Director*
Bonnie S. Lawrence, *Research Editor*
Victoria Cooper,
Contributing Researcher

Carolinda E. Hill, K. M. Kostyal,
Contributing Editors

Carl Mehler, *Map Editor*
Joseph F. Ochlak, *Map Researcher*
Martin S. Walz, *Map Production*
Tibor G. Tóth, *Map Relief*

Richard S. Wain
Production Project Manager
Lewis R. Bassford
Production

Jennifer L. Burke, Karen Dufort Sligh,
Illustrations Assistants
Sandra F. Lotterman, *Editorial Assistant*
Elizabeth G. Jevons, Peggy J. Oxford
Staff Assistants

Manufacturing and Quality Management

George V. White, *Director*
John T. Dunn, *Associate Director*
Vincent P. Ryan, *Manager*

Bryan K. Knedler, *Indexer*

Library of Congress CIP data: page 196

Contents

March 13, 1990: Violent tornado tears through Hesston, Kansas, one of 59 spawned in a two-day outbreak.

What Are

**The face of fear:
Laguna Beach, California, 1993.**

Raging Forces?
by Ron Fisher

Spaceship Earth, we call it, but it's more like a beat-up old bus bumping along a rough and rutted road. It's more than four-and-a-half billion years old, after all. An engine, inclined to overheat, chugs and sputters beneath the hood. The engine also powers peripherals—like the heater and the air conditioner—and keeps warm lubricants flowing over pistons and cylinders. Occasionally, at an especially severe bump, a door falls off the bus, or a window cracks. Cargo strapped to the roof shifts and teeters. Winds whistle past the open windows, swirl in, and blow peoples' hats off. Water sloshes in the cooling system, and steam comes boiling out of the radiator. Now and then, there's a backfire, which alarms the passengers.

A sorry old bus, our earth, but it's the only one we've got. And it is lovely—painted in deep blue and wispy swirls of cloud-white. But sometimes it seems intent on shaking us off, burning us up, drowning us, or wafting us off the planet totally.

The earth got put together in such a way as to ensure its continued restlessness. A solar system's worth of whirling gas and dust—two components essential to cosmologists' explanations of how it all began—gradually coalesced into a whirling disk with the sun at its center. What the sun didn't use became planets, moons, and asteroids. The pull of gravity and the bombardment of space debris heated the interior of our forming planet and separated the rocky materials into layers. Gases released as the earth's interior formed became its atmosphere.

Scientists believe a heart of iron 1,500 miles in diameter beats at earth's core. A liquid layer 1,400 miles thick surrounds that core; and a mostly solid mantle of hot, heavy rock, with a maximum thickness of about 1,800 miles, covers the two inner cores. The never ending heat from the interior causes a ceaseless churning and flowing within the mantle.

Like a hard-boiled egg that's been dropped, earth has a hard, thin skin patterned with cracks. Tectonic plates—some large enough to carry a continent, others

Steve McCurry/MAGNUM; S. Franklin/SYGMA (opposite); Rick Rickman/MATRIX (preceding pages)

smaller—drift like snail's-pace bumper cars across the face of the globe, sliding atop hotter, softer rock moving beneath them. The plates collide, they separate, they scrape past each other. They set off many of the forces that rage on the surface—the earthquakes, the volcanoes, the rising mountain ranges. Geologically speaking, they rule our lives.

Where plates spread apart, magma rises in the rift between them. There it hardens and moves aside to make way for more magma. This process built the 46,600-mile-long Mid-Ocean Ridge and the Great Rift in Africa.

Where an oceanic plate collides with another or with a continent, it often subducts, or thrusts beneath the other. A deep trench marks the spot. Where continental plates run into each other, they buckle, heaving up mountain ranges. Plates that slide past each other may create transform faults—like the notorious San Andreas Fault that so plagues California. Transform faults are one of the cradles of earthquakes.

Climatic Extremes

In Bombay (opposite) men hurry through a monsoon's deluge; at nearly the same latitude dust from a drought (above) envelopes a crowd in Sudan. Worldwide, in any given year, extreme weather events—such as heavy rains and long dry spells—occur. If extended, such events sometimes become floods or droughts. Monsoons, air circulations traditionally happening twice each year, affect perhaps half the world's population. From May through September, winds from the southwest bring heavy rains to most of southern Asia. In winter, the winds reverse, carrying rain to Indonesia and Australia. In Africa's Sahel, rainfall comes only three months a year and produces a short growing season. Any reduction is life threatening. Weather events may be called "raging" if they adversely affect people.

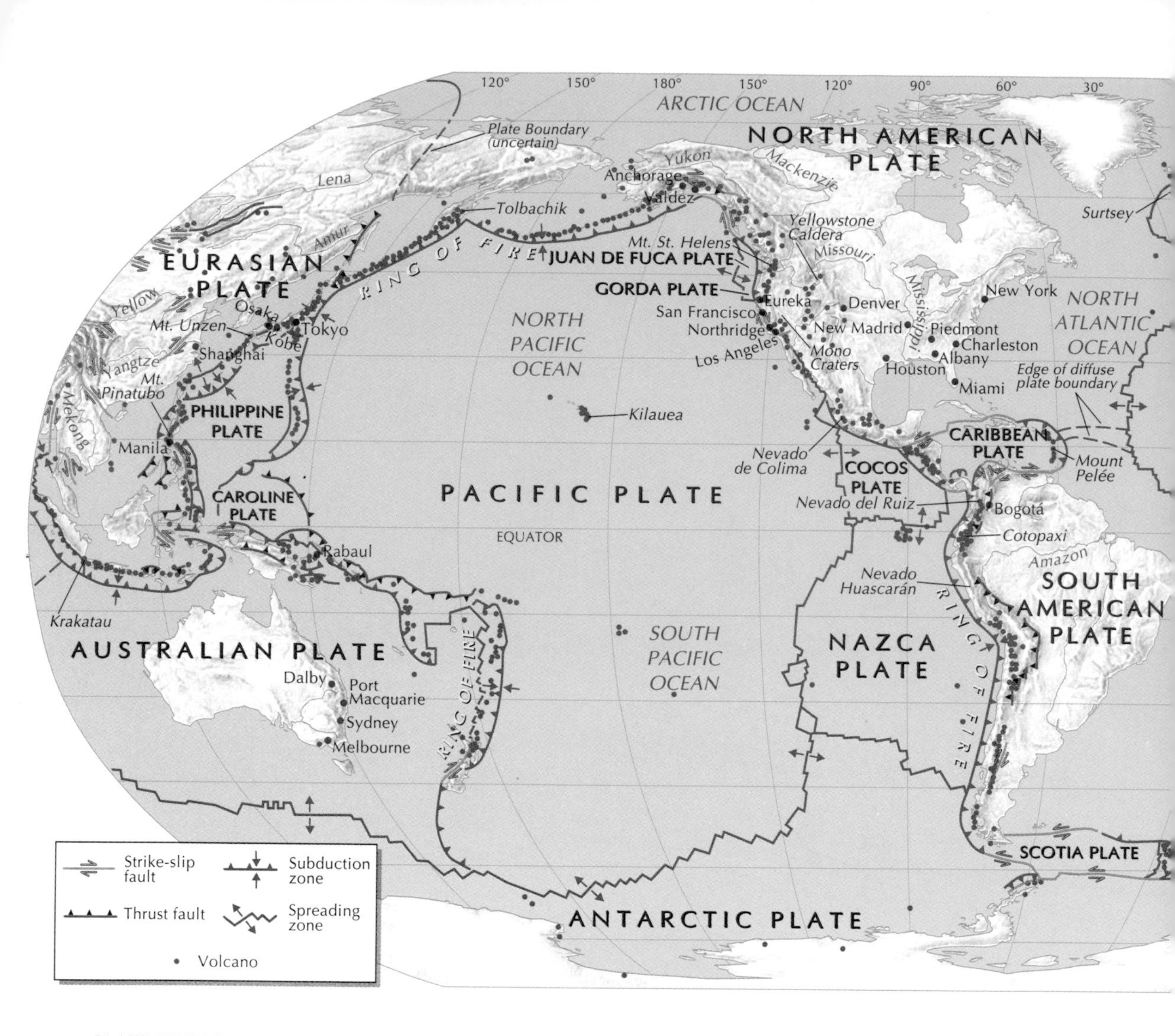

PLATE TECTONICS (ABOVE). The 16 large and several small plates shown on this map are thick slabs of rock on the surface of the earth. Their movement and interaction is called plate tectonics. Along their boundaries, they pull apart, collide, or slide past or beneath one another, triggering dramatic forces: earthquakes and volcanoes.

HURRICANES AND TORNADOES (BELOW). This map shows where hurricanes and tornadoes usually happen throughout the world. Hurricanes are born over tropical ocean waters; then they drift with the trade winds and curve toward the Poles. Colliding air masses give rise to tornadoes—perhaps 1,000 a year in the U.S.

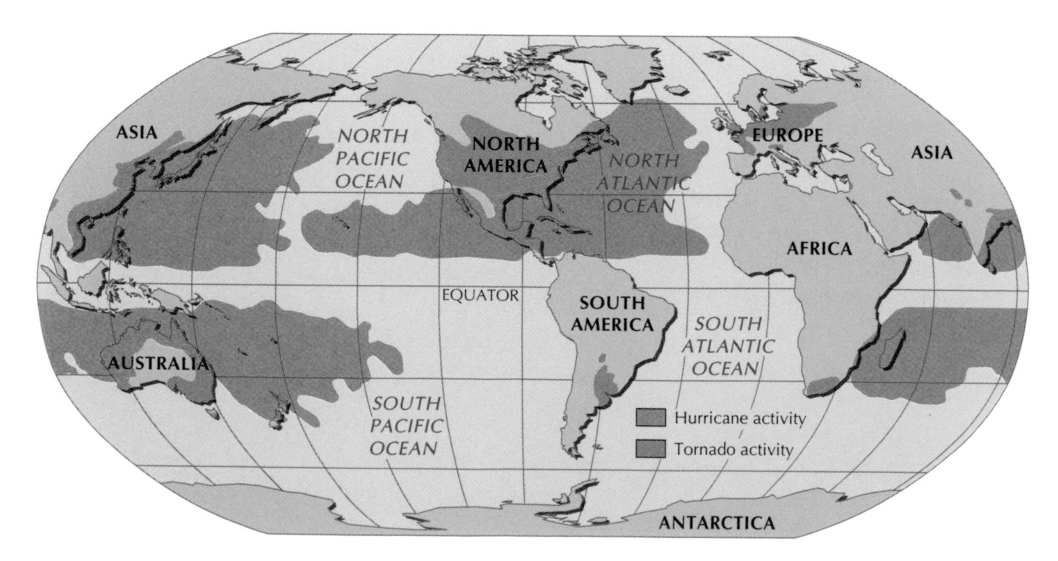

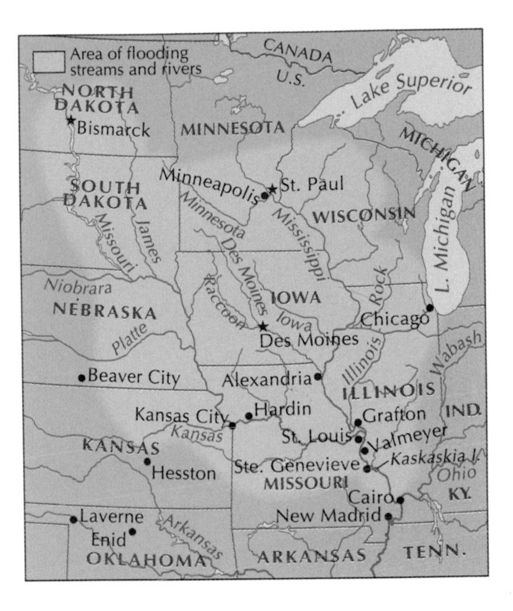

THE GREAT FLOOD OF 1993 (ABOVE). The zone of destruction and misery shown here affected more than 13 million acres in 9 midwestern states. Stalled weather systems brought record rainfall from April through August—more than 3 feet in parts of Iowa, Kansas, and Missouri—saturating the land and causing 100 rivers to overflow.

EL NIÑO (BELOW). Spanish for "the Child," El Niño is a periodic warming of tropical waters in the Pacific, causing westerly winds along the Equator, warm currents off South America, and weather anomalies around the globe. In normal years, trade winds blow steadily from east to west, dragging warm surface water in the same direction.

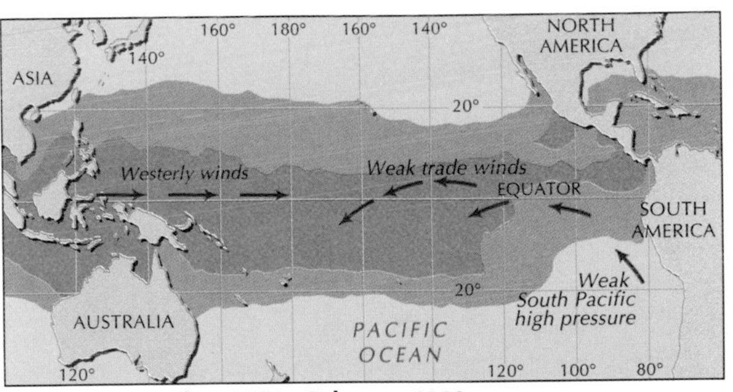

El Niño, December 1982—February 1983

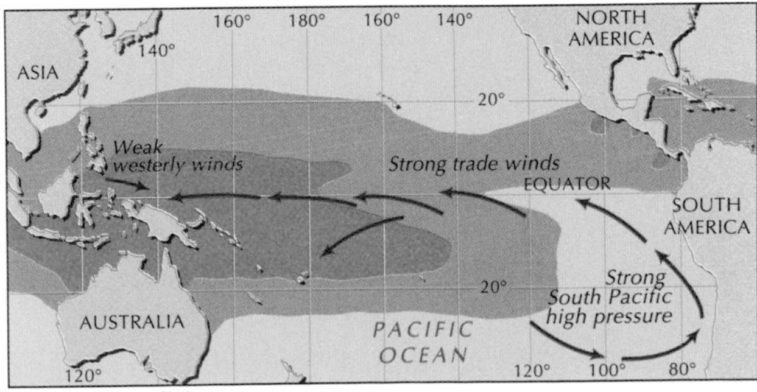

Normal Conditions

Our Planet Is Alive

Earth—nurturer of life—extracts a price for her bounty. The violent storms, seismic shocks, volcanic eruptions, extremes of weather, and swings of climate fascinate people at the same time they take a heavy toll of hardship, death, and destruction. Scientists work to understand the causes of nature's upheavals and try to reduce the impacts of their hazards and disasters. In the meantime, no place on earth is safe from the raging powers of nature.

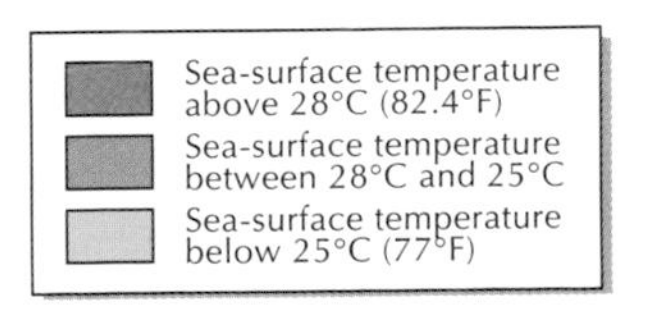

Most earthquakes occur near plate boundaries. While the plates slide freely, no problem. But when they get stuck, then suddenly lurch apart, walls tumble down and headlines get written. Volcanoes mostly rise where plates are either subducting or spreading. The edge of the gigantic Pacific plate—the 30,000-mile-long Ring of Fire—is especially active: About 70 percent of the world's more than 500 active volcanoes lie along it.

Brushing the coast of California as it does, the Ring of Fire simmers up and down the West Coast of the United States—especially when it comes to earthquakes. At more than 10,000 a year, earthquakes in California are virtually constant. Most are too slight to be felt by humans, but the needles on seismographs register their activity. And the swarms of aftershocks following large earthquakes keep epicenter regions agitated for days.

Forces work continuously to wear earth down. Other forces build it up again. The magma of the heated interior, ceaselessly searching for routes to the surface, seeps

Raging Earth Forces

Mount Whitney lofts its dramatic presence above the Sierra Nevada. The geologic forces that built Whitney also pattern California with a complex and far-reaching system of faults—which sometimes shift. Earthquakes result. A large quake that hit California in June 1992, emanating from the isolated Mojave Desert near Landers, hopped from fault to fault, traveling 43 miles in 24 seconds. It opened a rupture (below) and displaced a highway (lower).

Robert A. Eplett/CALIFORNIA GOVERNOR'S OFFICE OF EMERGENCY SERVICES (upper); Roger Ressmeyer/CORBIS (lower); Eugene Fisher (opposite)

through any crack or crevice.

The small island nation of Iceland straddles the Mid-Atlantic Ridge and is one of the world's premier laboratories for witnessing the formation, through volcanism, of new land. At least 20 eruptions at 8 sites in and around Iceland have occurred in this century.

The world watched an island being born there in 1963 when Surtsey, off Iceland's coast, steamed and thundered and rose from the sea. A fishing boat 20 miles south of Iceland was the first to notice: a sudden turbulence in the water. The eruption lasted for three and a half years, building a square-mile island. It's shrinking now, as erosion takes effect. Visitors to the island must empty their pockets, shoes, and trouser cuffs to prevent accidental importation of seeds.

Even without help from visitors, the first plant communities have developed. Sea sandwort appeared intermittently on the island for several years before finally producing seeds and becoming a permanent resident. Various gulls use a patch for nesting. Their excrement fertilizes the plants, an example of the symbiosis ever present in the natural world.

Another eruption, beginning in January 1973, brought big changes to the island of Heimaey off Iceland's south coast. A new volcano there erupted for more than five months, nearly burying the island's only town, a thousand-year-old fishing community, and threatening to close the harbor.

Within six hours after the beginning of the eruption, virtually all of Heimaey's 5,300 residents had been evacuated. But the decision was made to fight the volcano; around the clock, seawater was sprayed onto the advancing lava. Ultimately successful, this program was the most ambitious ever attempted by humankind to control volcanic activity. Between February and July some eight million cubic yards of seawater were pumped onto the creeping lava flow, converting about five million cubic yards of molten lava into hard rock.

Not satisfied merely to have defeated the lava, the ingenious Icelanders took steps to make it an ally. They built atop the mountain of cooling lava a heating system that, for about ten years, heated virtually every home on the island. "They pump down cold water and it comes up hot," an islander told me during a visit I made there in the early 1980s. Complex yet simple, the system relied on heat exchangers built over pipes buried above the lava, through which water percolated. Tephra was used to extend the runways at the island's airfield. The eruption added nearly a square mile to the island, an increase of nearly 20 percent. And the harbor, now with a lava breakwater, is actually improved.

In Hawaii, too, volcanoes are building new land. Kilauea, on the Big Island of Hawaii, is perhaps the most studied volcano in the world, with the Hawaiian Volcano Observatory on its flank since 1912. The volcano has been under nearly continuous eruption since January 1983 and by 1995 had added some 500 acres of new land to the island.

These forces, which have been active and natural since the planet's birth, become "disasters" only when humans get in the way. And, recently, that seems to be happening more and more. Earthquakes in California and Japan bring home to us the devastation a serious temblor can cause. Scientists estimate that there are some 500,000 detectable quakes scattered around the world each year. Of those, 100,000 can be felt by humans, and 1,000 are capable of causing damage. The strongest ever recorded in the United States was the Good Friday quake of 1964 in Alaska that killed 131 people and did some 750 million dollars in damage.

Earthquakes are among the deadliest of nature's cataclysms. A quake near the eastern end of the Mediterranean Sea sometime around 1201 may have killed more than a million people. Authorities agree that a quake in China in 1556 caused some 830,000 casualties.

The great San Francisco quake of 1906 pales by comparison. According to U.S. Army records only 498 people lost their lives in the quake and in the fire afterward. New scholarship, however, estimates the fatalities to have been at more than 3,000.

Robert A. Eplett/CALIFORNIA GOVERNOR'S OFFICE OF EMERGENCY SERVICES

Warning Center

Computer monitors glow, communications equipment hums, and warning controllers Sandy Green and Tom Flynn receive and disseminate disaster information in the warning center of the California Governor's Office of Emergency Services. They alert the state's officials to impending or occurring calamities—a busy job in disaster-prone California. The computer at right selectively records tremors around the state. Earthquakes shake California continually. Seismographs there detect more than 10,000 quakes a year, most too small to be felt by humans.

Most earthquakes occur where stresses along the plates build up until something gives. Seismic waves from a quake travel outward like ripples in a pond. There are three kinds: P, for Primary, waves are the fastest; somewhat slower are S, or Secondary, waves; the slowest are the Love and Rayleigh waves, which move along the earth's surface and cause most of the damage. Much of what scientists know of the interior makeup of the globe comes from studying the different ways the P and S waves pierce the earth.

Volcanoes, too, tend to lie along the boundaries between plates. Indonesia has the greatest concentration, with some 200 active volcanoes. The greatest volcanic explosion in modern times was there, the eruption of Krakatau in 1883. More than 160 villages were destroyed, and 36,000 people drowned in its companion tsunami

Robert A. Eplett/CALIFORNIA GOVERNOR'S OFFICE OF EMERGENCY SERVICES

(Japanese for "great harbor wave"). Dust from the eruption fell 3,000 miles away ten days later. The explosion may have been the loudest sound ever heard on earth.

Some volcanoes develop at subduction zones, where one plate slides beneath the edge of another. Heat melts descending rock, forming magma, which rises and erupts, building up volcanoes. Other volcanoes form where plates move away from each other, especially along the Mid-Ocean Ridge. Molten rock rises up and fills the gap between plates, then hardens into seafloor, which is carried away from the spreading center. Such peaks can eventually rise above the surface of the ocean, like Surtsey. At a few places around the world, magma forces its way through cracks in the middle of a plate. Eruptions there gradually build from their base hundreds of miles within the

Fallen Bridge

Like rockets, support pillars rise through the roadbed of their collapsed bridge, part of California State Route 1 near Watsonville. About ten miles from the epicenter of a 1989 earthquake, much of the city's downtown area was destroyed.

earth, and as the plate slowly moves, a chain of volcanoes trails along behind. The Hawaiian Islands are such a chain.

When a volcano erupts—perhaps after decades of quiescence—it can be devastating to humankind. In 1985, after nearly 150 years of slumber, Nevado del Ruiz west of Bogotá, Colombia, erupted. It was a tiny eruption, about three percent the size of the 1980 Mount St. Helens eruption, but it unleashed sufficient heat to melt some snow and ice on the summit. A flood of mud and water the consistency of wet cement—called a lahar—swept down the mountain slopes toward the city of Armero, whose 25,000 citizens were preparing for bed. The next morning more than 20,000 of them were dead, buried in mudflows after ten feet of volcanic debris had swept over their homes.

I remember the horror of being there a few days later: the blackened corpses emerging from the ooze, the stench, the devastation—and in the eyes of survivors, the awful resignation and fear.

Volcanoes threaten humanity in other ways. Lava can overrun homes, as it does occasionally in Hawaii. Eruption plumes can pose a hazard to aircraft: Volcanic ash can melt onto turbine blades, causing the engines to stall. Ashfalls can cover towns in fine powder. Volcanic fumes can kill plants. Pyroclastic flows or surges—mixtures of hot rock fragments and gases—can sweep down valleys and travel scores of miles beyond a volcano, destroying everything in their paths.

Predicting volcanic eruptions is a science that is still evolving. Scientists are caught in a bind between the need to warn people of impending events and the relatively infrequent activity of a single volcano. A volcano quiet for 600 or 5,000 years can suddenly roar to life in days or weeks. Scientists are improving techniques and equipment that will give reliable alerts. They continuously monitor

hot spots that show indications of activity.

For example, Yellowstone National Park lies atop a volcanic caldera. Although the area has not had a volcanic eruption for 70,000 years, something is clearly happening there now. The center of the caldera, which collapsed during a gigantic eruption 650,000 years ago, rose more than three feet between 1923 and 1985 and then sank nearly six inches between 1985 and 1993. One geologist said, "I think it's breathing."

Also potentially damaging is Mount Rainier, which looms just 40 miles from Tacoma, Washington, and 60 miles from Seattle. Though it hasn't erupted in about 150 years, there is good reason to believe that it will erupt again, geologists say. The greatest threat is from a large debris flow, a churning mass of water, rock, and mud spilling rapidly down a volcano's steep flanks. Mount Rainier's upper flanks and summit are under constant attack from hot, acidic, volcanic gases that break down solid rock into masses of unstable clay. Also, Mount Rainier has about 150 billion cubic feet of snow and ice on its upper flanks that could be melted by the heat of an eruption. A debris flow could surge down into valleys where more than 100,000 people live. Several hundred earthquakes have

When Earth's Tectonic Plates Shift

Villagers share the heart-breaking task of carrying a corpse to a funeral pyre, one of hundreds of pyres needed for fatalities from a quake that struck central India in the early hours of September 30, 1993. Mud-brick houses collapsed by the thousands, including 80 percent of the homes in Killari and neighboring villages, whose populations totaled 40,000. More than 9,500 people died. Survivors gave what comfort they could to grief-stricken relatives and neighbors (opposite). The temblor originated not at the boundaries of plates, where most quakes appear, but in a flat plain in the center of a plate. Similarly, rare "intraplate" earthquakes struck in 1811 and 1812 near New Madrid, Missouri.

Mark Simon/BLACK STAR; Swapan Perekh/BLACK STAR (opposite);Chris Johns (following pages)

Tectonics made visible: In the East African Rift System dormant Mount Longonot, northwest of Nairobi, Kenya, rises where lava flowed from deep rifts in the valley.

occurred around Mount Rainier in the past ten years, a sign that the volcano is far from dead.

I grew up in Iowa—no earthquakes, no volcanoes. We experienced only the occasional tornado, blizzard, or flood. The earth, when we thought about it at all, was simply the soil beneath our feet; it produced the crops that supported us. We were as far from harm's way as it is possible to get. Or so we thought.

In fact, the midwestern and eastern parts of the U.S. are vulnerable to earthquakes; they have, in fact, experienced countless quakes over the centuries. In 1811-12, three huge quakes centered near the little town of New Madrid, Missouri, caused great destruction over a 5,100-square-mile area. The shocks were felt over nearly two million square miles. Church bells rang as far away as Charleston, South Carolina. The San

African Eruption

Like fiery blood from a wounded planet, molten lava pours from fissure vents down the flank of Nyamuragira volcano in the Virunga Mountains of Zaire in May 1989. The Great Rift provides geologists with a laboratory for studying the elemental forces of uplift, volcanism, and faulting. Valleys in the Great Rift System form and subside as the earth's crust splits and the parallel rifts move farther apart. Eruptions are a by-product of the rifting. Some 30 million years from now, many geologists believe, the Somali plate may drift entirely away from the African continent, as Madagascar did 165 million years ago.

Chris Johns

Francisco quake of 1906, by contrast, was felt over an area of about 60,000 square miles.

If a quake the size of the 1811-12 one hit the New Madrid area today, it would cause extensive destruction in the central U.S. About 33 million people live in the seven states closest to the epicenter. Memphis, St. Louis, and Nashville could be badly damaged.

In 1886 Charleston, South Carolina, experienced an earthquake that caused substantial damage to about 2,000 buildings in the city. More than 100 of them had to be pulled down. Over 500 miles away, buildings rattled in Washington, D.C., and a trolley jumped its track.

During the last 2,000 years, large earthquakes may have occurred in South Carolina about every 500 to 600 years. And as there have been quakes in the past, there will certainly be more quakes in the future, geologists say.

Quakes in the eastern United States are less frequent and harder to track than those in the western part of the country, because the faults that cause them are generally deeper, in tectonic structures that are virtually impossible to find. Eastern quakes might cause more damage and loss of life than those in the West because they can produce damage over larger areas, because the East is more densely populated, because buildings in the East have been designed with little thought to earthquakes, and because earthquake awareness and preparedness are lower in the East.

Even New England is not safe. In November 1755 a large quake centered off Cape Ann in Massachusetts toppled chimneys in Boston. The weather vane fell off Faneuil Hall.

New York City, with its densely packed millions, seems especially vulnerable to an earthquake. Even under normal conditions, the city's aged water mains are routinely rupturing. Skyscrapers, built to flex in heavy winds, might survive a quake, but their acres of glass possibly would shatter and come crashing down on pedestrians. Decorative stonework, though beautiful, is vulnerable, as are the roofs of many large arenas and concert halls, which are supported by widely spaced columns. Buildings in older residential neighborhoods, constructed of unreinforced brick masonry, are at risk, though row houses with interconnected structures are more secure. Many newer buildings rest on filled marshland and reclaimed waterfront—land that can become less than terra firma in a quake. The huge, graceful bridges of New York, mostly suspension, are designed to sway, improving their chances of surviving a quake. Nobody is certain what would happen to highways and subways, built over a period of many years, with various construction methods and materials.

In 1995, the New York City Council approved new standards for construction, requiring reinforced masonry and stronger standards for steel and reinforced concrete, better soil studies, tighter regulations for building on landfill, and consideration of the effect of vibration on adjacent buildings.

New Yorkers have become complacent about hurricanes, too. A storm in 1938 that battered New England, killing 600 people, still ranks as the fourth deadliest hurricane in U.S. history and the most disastrous ever recorded in the Northeast. Today, after more than half a century of development and population growth along the Northeast coast, the potential for a far greater disaster exists. If a category-3 hurricane—with winds between 111 and 130 miles an hour—were to bear inland across central New Jersey on a northwesterly track, its more powerful right side would strike the metropolitan area squarely, raising a storm surge 20 feet high in New York Harbor. A storm of that track and intensity would cause damage far greater than that caused by Andrew in 1992. That same storm would raise a surge of 22-24 feet at Kennedy International Airport.

A mixture of heat and movement powers the weather machine responsible for such storms. Broadly, heat rising from the warmest parts of earth—those nearest the Equator—swirls as wind patterns around the globe, because of earth's spinning. The winds are twisted by the Coriolis effect, causing the ceaseless passage of the highs and lows that bring about weather changes in the mid-latitudes.

In the monsoon belt, which crosses the middle of the globe, there are alternating seasons of wet and dry as global heat exchanges move warm air to colder regions and vice versa. El Niño—the Child, so named for its usual appearance around Christmastime—results when Pacific trade winds falter. In 1982-83, an El Niño caused a set of some of the biggest natural disasters in recent times—droughts on four continents, major flooding on two, and ocean warming that disrupted fisheries along the west coast of South America. NASA scientists believe they discovered in 1994 a remnant wave from that El Niño, ten years after its birth, that still causes higher sea-surface temperatures in the northwest Pacific Ocean and appears to be affecting weather patterns. A few scientists have even suggested that this wave might have created the storm systems that caused the Mississippi Basin flooding of 1993.

An El Niño that began in 1991 continued into 1995, according to some scientists. They claim it was the longest El Niño ever. Others have suggested that it was, in fact, more than one event. It was blamed for flooding in California, for a rough 1994 winter and a mild 1995 winter in the northeastern U.S., for Hurricane Andrew, and even for the spread of cholera in South America.

Roger Ressmeyer/CORBIS

Hurricanes are defined as tropical storms with winds of at least 74 miles an hour. They are similar to the cyclones of the Bay of Bengal and the northern Indian Ocean and to typhoons in the western Pacific. Most Atlantic hurricanes are born off the west coast of Africa, where trade winds of the Northern and Southern Hemispheres meet. Energy supplied by warm water can cause a tropical disturbance to grow into a tropical depression, which sets off across the ocean in a westerly direction, often taking aim at the Caribbean or the East Coast of the U.S.

The waters of a hurricane, not the winds, cause the most deaths and destruction when a big storm sweeps ashore. Surges 25 feet high can inundate a coastline. The greatest natural disaster in U.S. history occurred in 1900, when a surge swept across Galveston, Texas, reducing the city to what someone called a crescent-shaped pile of rubble and drowning more than 6,000 people. An average hurricane drops more than 2.4 trillion gallons of rain each day of its existence.

And then there are tornadoes. Tornadoes have more concentrated destructive power than any other kind of storm, with

Geothermal Heat

Even in winter, bathers enjoy the warm, soothing waters of the Svartsengi geothermal plant in Iceland. The Mid-Atlantic Ridge extends above the surface of the ocean here. Home to some 800 geysers and boiling springs, tiny Iceland, a mountaintop in this chain, continually steams. The volcanic activity heats groundwater, springs, and water pipes, which Icelanders use to warm greenhouses, schools, homes, and pools.

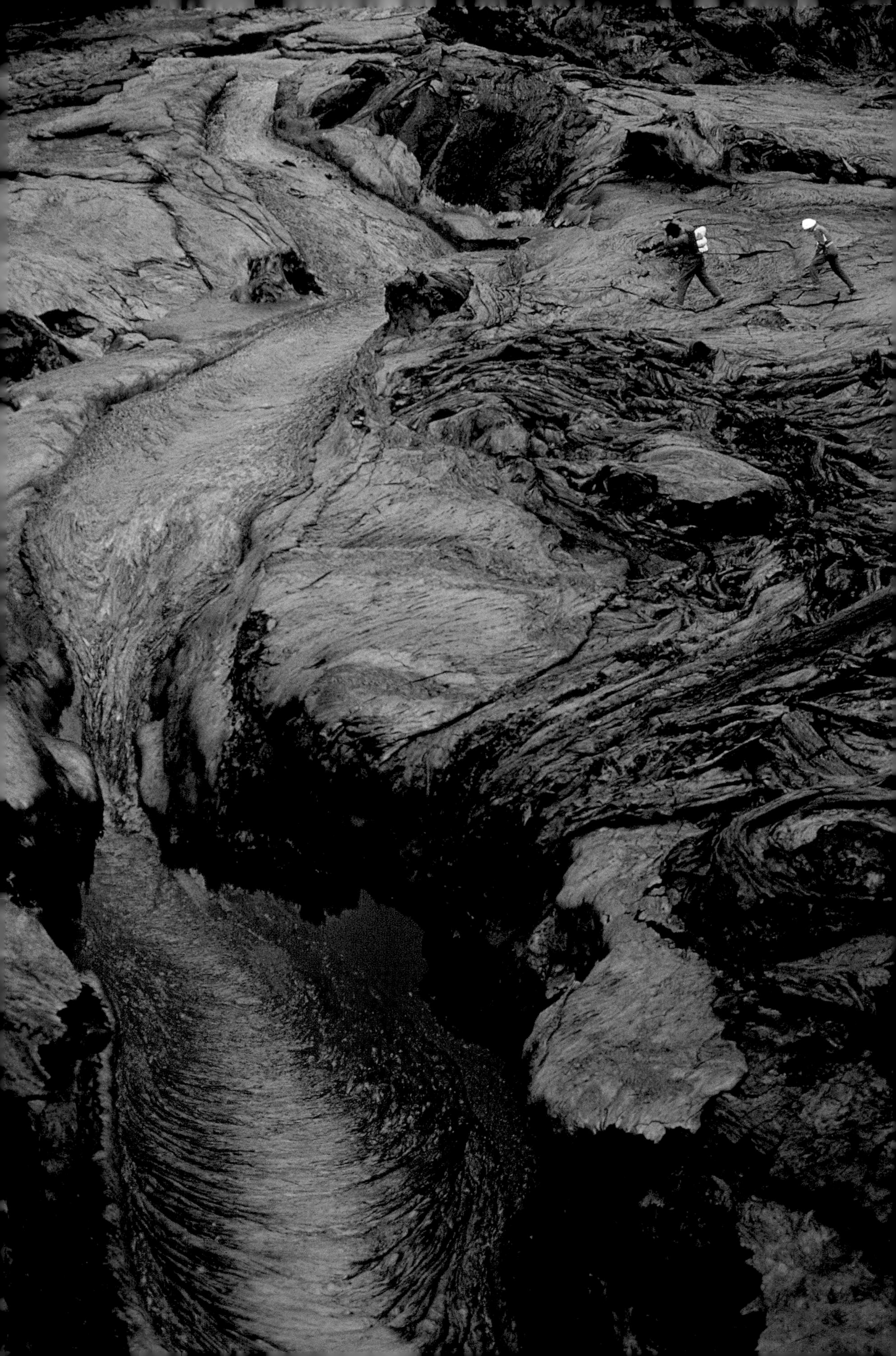

Hawaii—A Geologic Hot Spot

Geologists approach a river of molten lava flowing from Kilauea in Hawaii (opposite). Since 1983 the volcano has added some 500 acres to this island. But new land comes at a price. For example, of the destroyed community of Kalapana, little remains but remnants, such as the top of a school bus (below). A diver 70 feet deep finds pillow lava, which forms when lava vents into cold seawater. Hawaii sits atop a hot spot—a place where magma rises from deep inside the mantle. The hot spot stays in one place as the plate moves over it. The process results in a chain of islands.

Roger Ressmeyer/CORBIS (all)

Roger Ressmeyer/CORBIS (both)

winds that may reach 300 miles an hour. The strongest of tornado winds have never been measured, however, because no anemometer has yet been built that can survive them. Tornadoes are violently rotating columns of air that touch the ground during severe thunderstorms. The average one is about 660 feet wide, travels at some 30 miles an hour, and seldom moves along the ground for more than 5 or 6 miles.

More tornadoes form in the skies over the United States than in any other part of the world—between 700 and 1,000 each year. The so-called Tornado Alley where they occur most frequently, but not exclusively, stretches from north-central Texas through Oklahoma and into Kansas, from whence hapless Dorothy and Toto got lofted to Oz. The deadliest tornado on record killed nearly 700 people in Missouri, Illinois, and Indiana in 1925.

Tornadoes are born when cold, dry air and warm, moist air collide, producing a strong updraft. May is the month with the most tornadoes, but April's are the most violent. Across Tornado Alley, they fall from the sky like bombs—and cause just

Volcanic Giants Sleep

Looming in the distance (below), Mount Rainier gives thoughtful Washingtonians pause: Gigantic landslides 5,000 years ago from this volcano, still active today, sent mudflows over land now settled by hundreds of thousands of people. The flows inundated the Puget Sound lowland on the outskirts of what is now the Seattle–Tacoma area. More than a dozen potentially hazardous volcanoes form the Cascade Range, including Mount St. Helens (opposite, foreground), Mount Rainier (opposite, upper left), and Mount Adams.

as much terror and destruction.

But winds are a force to be reckoned with, quite apart from those of tornado intensity. An enormous variety blow, with many local labels around the world. Just to name a few: the chinooks of the eastern slopes of the Rocky Mountains, the brickfielders of central and southern Australia, the siroccos of the Sahara and the Mediterranean coast. Then there is the Santa Ana.... When a Santa Ana sweeps westward over the deserts and mountains and blows across southern California, it loses its moisture and warms up. Then it heats and dries everything in its path. People who live on the steep, chaparral-covered *(Continued on page 41)*

When Ill Winds Blow

Turning her back on a terrifying sight, Audra Thomas has her picture taken by her mother. The Thomases, who live near Beaver City, Nebraska, watched for "what seemed an eternity—probably 20 to 30 minutes"—on April 23, 1989, as seven or eight tornadoes formed, whirled, then dissipated or moved on. This one was moving away. Of earth's destructive winds, those making up tornadoes are the strongest, but hurricane winds last longer and hit a bigger area. From a satellite 1992's Hurricane Andrew (above) looks benign, even beautiful. But after the storm hammered South Florida, stunned homeowners could only stand and stare.

Lannis Waters/*PALM BEACH POST*; Rob Downey/Hank Brandli (upper); Merrilee K. Thomas (opposite); George Wilhelm/*LOS ANGELES TIMES* (following pages)

Whipped by hot Santa Ana winds from the deserts, raging fires in 1993 make parts of southern California an inferno.

P.F. Bentley/BLACK STAR;
Robert A. Eplett/CALIFORNIA GOVERNOR'S OFFICE OF EMERGENCY SERVICES (below and following pages)

California Fires Invade the Urban Landscape

In Calaveras County, California, in August 1992 a firefighter braves the Old Gulch conflagration. The fire burned 17,000 acres and destroyed 170 structures. Another killer fire advances inexorably through the hills of Oakland in October 1991 (opposite). The state's combination of ingredients—periodic dry spells, dense vegetation, high winds, heavily populated hills—frequently leads to costly and devastating fires. The Oakland fire blazed across 1,500 acres. It killed 25 people, destroyed some 3,000 homes, and burned 26 miles of city streets.

In 1993 a home survived the Laguna Canyon fire that burned 366 structures near Los Angeles.

STOP
STOP

Steve Starr/SABA; AP/WIDE WORLD PHOTOS (opposite)

hillsides hold their collective breath.

Perhaps Californians are cursed by beauty. Their state attracts settlers to its most beautiful parts, which are also the most dangerous. Sixty-one percent of California is covered with wildlands; fire there is a natural part of the ecological balance. As the state's population grows, more and more people move inland to communities that border on wildlands.

"This means that more people are in peril. And since 90 percent of our fires are caused by people, the wildlands, too, are more imperiled," says a spokesperson for the California Department of Forestry.

The worst urban forest fire in United States history occurred in California in 1991, when the picturesque Oakland and Berkeley hills across the bay from San Francisco burned. Twenty-five people were killed, and 3,000 homes were destroyed. Conditions were right for a blaze. Temperatures in the area had reached into the 80s and 90s, unusual for October. On the day the fire started,

Floods and Slides

Beleaguered Pacific Coast Highway suffers yet another crisis: Heavy rains early in 1995 bring traffic to a halt and heavy equipment to the scene. Here in Malibu, scenic State Route 1 threads between coast and mountains, providing access to expensive beachfront property. But landslides and mudslides frequently bring unstable material from the mountains onto the road; it was closed 300 times in the last decade. Earthmovers (opposite) clear a path through debris covering several hundred yards of highway. Streams that emerge from the mountains here carry not just water but also tons of boulders and other debris. Above, a backloader, hurrying to stay ahead of the flood, scoops rubble that threatens a highway bridge across Las Flores Creek. The plant nursery's floor and creekside walls collapsed.

humidity early in the morning was unusually low. Dry, gusty winds from the north—which firefighters dread—were blowing across Oakland.

It was later determined that the fire began in an area of low grass. How it started remains a mystery, but it quickly consumed nearby dry trees and brush. Gusty winds blew it to the southeast, through residential neighborhoods. Houses became the fuel. Burning cedar shingles, torn loose and carried aloft by the wind, carried the blaze quickly across streets and highways. The combination of high winds and burning embers spread it across 1,500 acres of homes and businesses—homes with an average value of $600,000. Coming on the heels of a 1989 earthquake that did its damage in the city's poorest area, this fire devoured the lush hillsides of the city's wealthiest.

Months later, the *New York Times* reported that crocuses were poking through the rubble and ash, "...along with stubborn crabgrass and dandelions.... The ravaged neighborhoods are strangely beautiful.... twisted patio furniture, rusted husks of water heaters, and stone stairs leading nowhere stand like bold works of sculpture.... Foundations on a distant slope laid bare by the fire have the patchwork geometry of farm fields seen from an airplane."

On October 19, the Saturday before the fire, Fred Booker, a graduate student in geology at the nearby Berkeley campus of the University of California, suspected there was a fire somewhere. A helicopter kept flying back and forth over his house in the Oakland hills, collecting water from a small, man-made lake nearby. "The next morning," he told me, "I was in my backyard watering plants. The Santa Ana type winds were blowing out of the north, very hot and dry. I was just thinking to myself, 'Good thing the fire was yesterday.' A few minutes later I saw wisps of smoke coming over the ridge."

He went inside the house and warned his wife that it might be a good idea to gather up important papers, in case they had to leave. "I went back outside and walked down the street to get a better feel for what was happening."

Suddenly a grove of eucalyptus trees burst into flame. Just erupted. He hurried back to his home. "My wife was still in her caftan, talking by telephone with her cousin. She had the vacuum cleaner out and was cleaning up the fine ash coming in through the window. She was thinking, 'Well, if there's a fire, they'll put it out.' I told her, 'Get dressed. We've got to leave.' She said, 'You're kidding.' I said, 'Nope.'"

Fred went back outside and walked the street again, to make certain neighbors were aware of what was happening. In five minutes he was back. A Monterey pine behind his house was on fire. "My hose was still out, so I grabbed it to spray down the tree, but there was no pressure." His wife was ready to go, and they walked out the door. "I noticed some neighbors were still at home. I went to see what the problem was. The wife wouldn't leave because she couldn't catch her cats. They were panicked. I grabbed her, dragged her to the front door, and showed her a tree burning at the bottom of the steps. Then she understood. By then there was fire all around us. It was as black as night. As we drove out, I could not see the front of the car. Burning embers were blowing past us horizontally, in winds that must have been 30 to 40 miles an hour. Brush on both sides of the road was burning. My eyes were full of ash. It was a nightmare." All that survived of their home was the mailbox.

And a few days later, it began to rain.

A popular perception is that rains following a fire on a steep California slope mean one more misery: mudflows. But such is not always the case.

Fred Booker wrote his master's thesis

Salvaged

From someone's living room, possessions damaged, but not lost, poignantly survive in Roseville. This central California city was hit by flooding from nearby Linda Creek in January 1995. Nature's wrath constantly reminds humankind: The forces that make the earth a living planet also make it a dangerous place to live.

Randy Pench/SYGMA

on the impacts of the Oakland fire on hillslope erosion and hydrology. Immediately after the fire, officials spent almost five million dollars on an ambitious erosion control program, in which the primary mitigation effort was hydroseeding: Native grasses mixed with adhesives were sprayed on hillsides. Booker and his colleagues would find it was money that didn't need to be spent. This kind of control program prevents only surface erosion—and that, only if the grasses germinate and provide cover before the first heavy rains of the next winter.

In March of 1995 I met Fred at the Los Angeles airport, and we set off to see some of the places in southern California where earthquakes loosen slopes, fires burn the vegetation off them, gravity pulls rocks and sediment into waiting channels, and rainfall flushes this devil's soup onto people and their works.

John McPhee once wrote, "In millennia before Los Angeles settled its plain, the chaparral burned every thirty years or so, as the chaparral does now."

"Peoples' memories are short," said Fred, as we drove up the Pacific Coast Highway toward Malibu. "They think that because something hasn't happened yet, it's not going to happen." He nodded at the slopes along the highway. "That's chaparral. It's made up of vegetation that has a high oil content. When it burns, it produces a vapor that penetrates the ground, condensing as it cools, creating a waxy layer—called a hydrophobic layer—just beneath the surface. Essentially, it makes soils refuse to 'wet up'. The hotter the fire, the deeper the layer. All the material above a hydrophobic layer has the potential of being transported off the slope and into a channel and then into somebody's bedroom."

The Pacific Coast Highway has been closed more than 300 times in the Malibu area in the last 10 years, mostly because of landslides. At several places along the highway, workers with huge earthmoving equipment were shoring up streambeds, cutting back cliff faces, or repairing damage from the latest storm. In February of 1994, 33 houses here were surrounded by a sea of mud that entered more than half the homes. Mud piled up against another 13 houses.

We drove up Las Flores Canyon. In January 1995 flooding here closed the Pacific Coast Highway with mudslides; Las Flores Creek overflowed, necessitating use of a backhoe by emergency crews to remove debris; about 50 people had to be evacuated from homes and businesses; and the Las Flores Creek bridge supports were eroded, temporarily closing it.

We stopped beside the stream. A trash rack in the streambed was intended to stop debris washed out of the mountains during a rain from collecting at the undersized highway crossing and imperiling (or closing) the only link among communities along the Malibu Coast. Farther up the canyon we stopped at what Fred called "the mud house." It was being rebuilt. "Rain had awakened the owner. He got up and went into his kitchen to get something to eat, and just then the mud came in through his back wall and into the bedroom he had just left. The weight of the mud collapsed the floor and the mud flowed down into the garage, which was below the bedroom, and on out onto the street." The house was directly in the center of a small ravine. It was as if it had been built in the barrel of a shotgun.

As we drove up into the San Gabriel Mountains north of Pasadena, Fred talked with me about the nature of slope failure following fires. The San Gabriels are high mountains—up to 10,000 feet in altitude—and extremely rugged. They seem to go straight up from the neighborhoods and loom over them, beautiful but unstable. Every little shake or rainfall brings more of them down, as if they were consciously trying to dismantle themselves. They were scarred by fissures, where sections of them had simply let go.

In Altadena, Fred took me to a debris basin, a large, man-made bowl covering several acres, like a shallow bathtub. It was built to catch rocks, boulders, and mud coming off the mountains during storms. A dam stops the big debris, and an outlet tower lets the water continue on its way. The mountains of coastal California are sprinkled with these structures. Practically every canyon emerging from the San Gabriels has something similar waiting at its mouth to trap whatever comes down.

By late afternoon the tops of the San Gabriels had disappeared into misty black clouds, and rain drops were sliding down our windshield. It was raining.

Gerry Gropp/SIPA PRESS

Flooded Fields, Higher Prices

Strawberry fields in Pajaro, California, flood in 1995. This year was the first in which floods occurred in every county in the state. By mid-March 50,000 acres of farmland were under water. California's flood damages to agriculture from January through April caused losses of nearly 750 million dollars. Consequently, supermarket prices for lettuce, broccoli, artichokes, and strawberries rose throughout the nation.

When the Earth

A torch of lava bursts from the 1975
giant fissure eruption on the flank
of Tolbachik in Kamchatka, Russia.

Moves
by Noel Grove

The proud works of humanity lay jumbled like contents on a Monopoly board kicked in passing. In the pale glow of moonlight houses and hotels sat at odd angles, or on their sides, or collapsed in wreckage. Some, still standing, were bent in the middle. Sidewalks and streets bulged and gaped with huge cracks, and freeway overpasses had toppled sideways. Quaint old Japanese buildings of age-darkened wood lay in heaps, like kindling. Around campfires made of that "kindling" huddled small clumps of people, as if primitive nomads had wandered into the ruins of a once great city.

This was Kobe, Japan, as I saw it before dawn, four days after the earthquake-prone nation had been jolted by one of the most violent temblors in nearly three-quarters of a century. Those around the campfires were among the 300,000 left homeless when the ground shook for 20 seconds beneath Japan's sixth largest city at 5:46 a.m. on January 17, 1995.

In the Kobe quake, dubbed the Great Hanshin Earthquake by the media, 5,500 people died and 25,000 were injured. It had struck directly beneath the downtown area and had measured 6.9 on the moment-magnitude scale.

Fires fed by severed gas lines raged for days as firemen stood by helplessly, disarmed by broken water mains. Food, blankets, and medical supplies for the living were delayed by a crippled transportation system—bent rails, broken roads, cracked runways.

To reach Kobe and to avoid the worst of the traffic on the one remaining highway, I had left Osaka, 20 miles to the east, by car before daylight.

By a campfire in a street stood 44-year-old Yuki Ishii, a printer, whose apartment

Hashimoto/SYGMA; AP/WIDE WORLD PHOTOS (opposite); Vadim Gippenreiter/REI ADVENTURES (preceding pages)

in a nearby building was destroyed. "I was asleep when the bed started moving from one side of the room to the other, and bouncing up and down," he told me. "The building collapsed, but two walls tipped together and I escaped through the open space between them. The print shop where I work was badly damaged, so for now I have no work, no home, and no plans for the future. People bring me food and water." He held his hands to the warmth of the fire and sat on a chair cushioned in red velvet, salvaged from a destroyed coffee shop.

As morning grew brighter, I saw dozens of the newly homeless washing their clothes in a small stream. "Our apartment jumped up, down, and sideways, but it did not fall," said a woman squatting by the water with her daughter. "The building next door collapsed and three are dead. A neighbor is buried under rubble, her condition unknown."

Several people who had lost homes, friends, family members, and jobs in the quake shrugged and told me, "It was a natural occurrence; there is nothing we can do about it."

Nightmare in Kobe

In the early morning of January 17, 1995, Japan's most damaging temblor in 72 years, dubbed the Great Hanshin Earthquake, struck and paralyzed the city of Kobe. Records show that 5,500 people died. Survivors clustered around telephones (above) to tell relatives that they were alive. The injured and homeless numbered in the hundreds of thousands as Kobe homes, highways, factories, and port facilities were crunched. Severed gas lines started fires that burned for days (opposite). The awesome jolt measured 6.9 on the moment-magnitude seismic scale, twice the severity of the 6.7 quake that struck Northridge, California, a year earlier to the day. The moment-magnitude scale, now preferred by scientists over the older Richter, is a measurement of the total energy released by earthquakes.

Patrick Robert/SYGMA; ASAHI SHIMBUN (opposite)

"Japanese believe there are four powers that are...insurmountable...all-powerful...cannot be overcome," said Kunio Kadowaki, my guide and interpreter. "They are *jishin, kaminari, kaji,* and *oyaji*—earthquake, lightning, fire, and father." He chuckled. "In modern life father is becoming less than insurmountable. Earthquakes always head the list."

Not all of us are so mentally prepared, so reminded that we live on the thin exterior of a volatile, living planet. Beneath the hard skin of the earth lie miles of more pliable rock. Even deeper than that, extending into the planet's core, red-hot metal writhes and circulates.

The cooled, hardened ground on which we live floats over this inner, malleable cauldron as some 16 major plates and many smaller ones. Japan sits at a junction of major plates, as does the state of California.

Across the Pacific from each other, Japan and California form two boundaries of the Ring of Fire, the area around the Pacific Basin where the subduction of

Transportation Chaos

Passing directly under Kobe, the 1995 earthquake brought life in a thoroughly modern city to a standstill. Rails were twisted like spaghetti; the train cars on them were tossed about like toys (above). The Hanshin Expressway toppled on its side, spilling a delivery truck on early morning rounds and its cargo onto the street below (opposite). Japan's earthquake specialists had insisted earlier that well-built Japanese structures would weather temblors better than buildings in California. In reality, those in Japan erected after strict codes had been enacted fared well, but older structures and wooden houses with heavy tile roofs collapsed. Single-column supports under the expressway failed for lack of enough steel reinforcement. Broken roads and rails delayed help from nearby Osaka.

An Aftershock of Horror

Terrified by the events around her, an elderly woman cowers in a pickup truck. Older people were especially vulnerable, living in the fragile houses of an earlier time. With water unavailable, Kobe residents stare helplessly at the outbreak of flames that destroyed city blocks. The early hour of the quake, striking without warning along a minor fault, may have reduced the toll of victims; later in the day freeways, bullet trains, and sidewalks would have been filled with commuters.

Patrick Robert/SYGMA

plates causes frequent jarring and volcanism. As plates subduct, some of the solid material becomes molten again in the heat, rises as volcanoes, and erupts over the land as recycled rock.

The advance of technology over the past half century has given us a clearer picture of inner earth. By studying the passage of the waves of energy released by earthquakes through the planet, scientists have been able to map its deep structure, including a zone made of liquid iron-nickel alloy around its core. Seismometers detect and locate the vibrations of rock movement far beneath us.

Ground-based and satellite measurements detect the small surface movements (centimeters or less) of a volcano that typically occur before it erupts. Analysis of volcanic gases by spectrometers and other instruments sometimes foretell an eruption. Such volcanic-monitoring devices give us clues of possible cataclysmic events that may affect our lives, but we are powerless to prevent them.

Though not a scientist, I have witnessed as a journalist these forces and have been awed and occasionally terrified by them. Over the span of nearly a quarter of a century I have seen a dozen volcanoes erupting. I have climbed on and studied at least that many more as they sat dormant. One does not plan to experience an earthquake, but traveling often in areas of geologic instability has put me in touch with several of them, as well.

I am a moth sometimes foolishly drawn to the flame of inner earth. During a 1973 eruption in Iceland that nearly covered the village of Vestmannaeyjar, I approached so near the vent of the volcano one night that the blazing lava bombs arcing skyward fell behind me and the barely cooled lava oozing from the crater scorched the soles of my shoes. A few dozen yards away, jets of red-hot

三菱銀行

A City Scrambled

Both the modern and the vintage crumpled in Kobe's 20 seconds of intense vibration. A glass-fronted building sited on a filled estuary sagged and showered its windows into the street (opposite). Her home destroyed, 82-year-old Iyako Okubo carries a few retrieved belongings past leaning utility poles (above). An apartment building did not crumble but fell on its side (right). The landlord crawled out a window onto the street, where author Noel Grove found him praying for his tenants, several of whom had died. The author observed a remarkable spirit among the survivors. Cleanup and repairs began immediately; the atmosphere in shelters was upbeat. With automotive and rail traffic much hampered by quake damage, supplies poured in on the backs and bicycles of relatives and friends.

ASAHI SHIMBUN (above and opposite); AP/WIDE WORLD PHOTOS (above left)

Michael Yamashita; Philippe Bourseiller/HOA-QUI (following pages)

Havoc in the Harbor

No safe haven for vehicles, Kobe's harbor sucked in cars and vans when wharves collapsed. Japan's most important heavy cargo port was shaken into inactivity after the quake, but making repairs rapidly and shifting some port activity elsewhere brought exports back to normal by spring. Construction on filled estuaries had failed when the quake liquefied loose, wet soils. Deep-sunk pilings could have created solid footing. Despite extensive research, the temblor's severity surprised Japan, striking a minor fault of a complex tectonic zone that had been quiet for half a century.

lava burst forth spasmodically with a deafening roar. There is no sound on earth like that of earth clearing its throat and coughing new land upon old.

Eighteen years later, still trying to look down the planet's throat, I knelt at the rim of a grumbling volcano on the island of Espiritu Santo in the South Pacific and peered at the lava pool bubbling and dancing hundreds of feet below. I knew the pressure built up regularly so that the vent cleared itself with a thundering blast every 30 seconds or so. And I knew that in its current activity the leaping lava would not reach the rim where I perched. Yet, when it happened, the bottomless power of the explosion made me fall backward in fear.

Earthquakes do not hold that same attraction for me. When normally reliable terra firma begins trembling and turning my legs to jelly and causing buildings to sway like dolls' houses, my first feeling is that of betrayal. Once in the Philippines, when the house in which I slept was jolted by the second sharp tremor of the night, I was awakened by the sound of my own voice, yelling in terror.

These events bring the world back into proper focus. A current of narcissism seems to run through our advanced societies, fed by the assumption that our species now dominates the earth. Earthquakes jar our faith in our specialness. Volcanoes hurl ash and lava on our self-image. Both remind us that we are mere passengers on massive plates constantly on the move, grinding and gronking and leaking new materials, and we realize that for all our cleverness we have no more control over these journeys than do penguins on icebergs.

Since we cannot control earth's raging forces, we try to predict when they will happen. Science has inched closer to that goal. Volcanoes, especially, give clues that they may be building to a new outburst. One of the best examples of preventive volcanology so far came with the eruption of Mount Pinatubo. The Philippine Institute of Volcanology and Seismology, known as PHIVOLCS, set in motion a pre-explosion study that accurately predicted one of the largest eruptions in this century and, undoubtedly, saved many lives. Fewer than a thousand people died—some from the volcano itself and others from diseases contracted in refugee camps and from cascades of muddy material called lahars—loose ash washing off the volcanic highlands during heavy rains.

In early April of 1991 villagers living on the slopes of Pinatubo reported that the mountain was spouting steam and ash from several vents. PHIVOLCS began monitoring Pinatubo, which had not erupted for 600 years. Seismometers were installed on the slopes to detect the small earthquakes caused when magma is pushing its way up through formations of rock. Tiltmeters were set up to see if pressure from magma was causing the mountain to bulge.

The people of the Aeta tribe were in the most danger, but 15 miles from the mountain sat the scattered city of Angeles, population 236,000, and Clark Air Base, with some 20,000 United States servicemen and their dependents. The PHIVOLCS geologists were joined by a team from the U.S. Geological Survey, and together they began sampling soil and rock around the volcano for clues to how it had acted in the past.

"We found deep layers of pyroclastic flow material," said Raymundo Punongbayan, PHIVOLCS director, "indicating that Pinatubo eruptions tend to be very violent." Pyroclastic flows occur when magma becomes so energized that it turns into a surging red-hot cloud of gas and dustlike material.

As surface sampling continued, the seismometers were sending increasingly frenzied signals to monitors at Clark Air Base. The geologists urged evacuation. Ordering people to completely disrupt their lives, however, not only is costly but also can be politically embarrassing if nothing happens. When the geologists finally issued their highest level of alert on the tenth of June, Clark sent packing all but a skeleton crew of some 1,500 individuals. The Philippine government ordered evacuations of everyone within 12 miles around the volcano, and 200,000 people left their homes.

Explosions increased and earthquakes continued. At dawn on June 15, 1991, Pinatubo fulfilled all the geologic predictions, decapitating itself with a blast that removed the top thousand feet of the mountain and sent sulfur dioxide 10 miles high. Mingling with water vapor in the stratosphere, it formed a cloud that would girdle the earth. (The bitterly cold winters and cool, rainy summers in much of North America and elsewhere in 1992

After the 1991 eruption of Mount Pinatubo in the Philippines, residents near the volcano carry umbrellas to ward off showers of gritty ash.

and 1993 occurred in part because Pinatubo reduced the amount of sunlight reaching the earth by about two percent before the cloud eventually dissipated.)

More than two cubic miles of pulverized material sprayed over the countryside around the volcano. Soaked by the rains of Typhoon Yunya, which swept Luzon at the same time, the sodden, heavy ash collapsed houses, killing some residents. All that material exiting the earth left a giant cavern deep beneath the surface, and it was that cavern falling in upon itself that shook me awake and yelling near the volcano days later.

I had arrived in Manila the day before the major blast. After the eruption, I watched with astonishment in mid-afternoon as the city, 55 miles from the mountain, turned black as night from combined ashfall and rain.

Driving north toward the volcano the next day, I met a stream of evacuees in cars, on bicycles, on foot, or clinging to the sides of trucks. Trees were bent and broken from the weight of the ash; rice and sugarcane fields were covered with it. Power lines were down; houses lay in shambles. And still Pinatubo continued to rumble, less violently than before but continuing to send out pyroclastic blasts capable of scorching anyone within ten miles of the volcano. Heavy rains continued to wash down powdery ash from the heights in hot floods of cementlike slurries—the lahars that covered fields, highways, and even entire towns for miles downstream.

I joined Philippine and American volcanologists for a helicopter survey of the volcano and its surroundings. The caldera itself was obscured by steam and clouds, but the landscape around it reminded me of photographs of the moon's surface—gray and featureless. Entire valleys, 600 feet deep, had been filled to the ridges with the pyroclastic material. Tons of the

Philippe Bourseiller/HOA-QUI

gray ash sat poised in the highlands, awaiting more heavy rains that would carry the material downstream. "The threat of house-burying lahars will remain for at least five years," said Chris Newhall, who headed the USGS team.

And so it has. I returned to Pinatubo three and a half years later, weeks after the latest lahar had roared out of the mountains and buried parts of the towns of Porac and Bacolor, killing 28 people. Felix Yumang and Lila Andriguez led me

through a desert of now dry sand and gravel to what remained of their neighborhood in Porac. Their homes had sat alongside a pleasant stream, with fish for the taking and greenery lining the banks. Now the vegetation was gone and the once pleasant stream was a muddy trickle a yard wide and inches deep. Felix pointed to a foundation where his house once sat before the torrent carried it away. Lila pointed vaguely to a patch of bare earth where all trace of her home had been

A Land Transformed

Seeking greener pastures, farmers ride water buffalo through a gloom of falling ash. Besides disrupting Philippine life and landscapes, Pinatubo's violent outburst temporarily cooled the earth.

Living With the Violence of Volcanoes

Japan's unstable geology spawns eruptions such as this from Mount Unzen (above). Lava from a thick dome at its summit at times breaks loose and rolls down the mountain in fiery clouds called pyroclastic flows. One such flow killed 43 people, including renowned volcanologists Maurice and Katia Krafft. Heavy rainfall triggered rivers of mud that wiped out a subdivision at nearby Shimabara (below). Explosive Sakurajima has bombarded the communities at its foot for years, requiring safety training for students at Kurokami Elementary School (opposite).

Roger Ressmeyer/CORBIS (all)

wiped out or buried. "We were warned ten minutes before it came, around midnight," said Lila, "and escaped with the clothes on our backs."

The altering of entire landscapes became more apparent to me when I joined PHIVOLCS on a helicopter ride over the same terrain I had seen just after the eruption. The former moonscape was now deep-cut with ravines; on the slopes grew young banana trees and traces of other vegetation. I was told that deer had drifted back into the mountains and that some of the Aeta returned during the dry season, retreating again to refugee camps when rainstorms threatened new lahars.

For the first time, I saw the crater itself, a rock-rimmed hole a mile and a half wide forming a chalice for a lake colored milky green by sulfur. A sharp-peaked island had grown in the middle where lava had oozed out of the vent.

"Someday it will be a tourist attraction, if the lava doesn't fill the whole caldera," said PHIVOLCS' Punongbayan.

The helicopter followed one of the main drainages of lahar material downstream where we saw thousands of acres of croplands buried in new fans of volcanic debris, ruined for years. Cranes

THE COURIER-MAIL, Brisbane, Australia

and trucks were busy moving the dried material into dikes 18 to 20 feet high to direct the flow of new lahars, but the thick currents will still, in places, break through. Sometime in the future these lahar-covered lands will be fecund with crops again, for volcanic materials are rich in minerals important to plant growth, such as phosphorus, calcium, potassium, magnesium, and sulfur. In fact, the very atmosphere that allows plants and animals to survive resulted from volcanism's liberating moisture and gases from rocks.

Past contributions and future fertility are of little comfort to lives disrupted. Shortly after the eruption, 130,000 people lived in shelters, and more than three years later, some 47,377 remained.

Pinatubo signaled a victory for volcanologists, who showed they could predict a major eruption. Unfortunately, volcanoes can be filled with surprises. Weeks before Pinatubo erupted, a volcano in Japan called Unzen had been oozing a thick, viscous lava that gathered in huge clumps at its peak, occasionally tumbling down the sides in small pyroclastic flows. I still have the letter I received from a

A Leaking Planet

Twin eruptions spew ash near Rabaul, Papua New Guinea, marking both sides of the bay carved by a huge eruption 1,400 years before. Scientists worry that magma escaping from two locations may signal a weakening of the caldera rim and another large caldera-forming explosion. By monitoring signs of unrest such as earthquakes and ground movements, they correctly anticipated this 1994 event.

young geologist named Harry Glicken, who had heard that I was working on a magazine article about volcanoes. "Come to Shimabara, and I will act as your guide on Unzen, which is doing some interesting things," he wrote. I replied affirmatively and arrived in Shimabara, Japan, on June 8. Had I arrived five days earlier, I might have been a footnote in this book instead of one of its authors.

On June 3, 1991, Harry and two French volcanologists, Maurice and Katia Krafft, went up a valley on the side of Unzen to see the small pyroclastic flows that had been surging down the mountain as lava broke loose from the growing dome.

Films by the Kraffts had been shown worldwide and had contributed much to the public's understanding of the dangers of volcanoes. Lacking, the Kraffts felt, was dramatic depiction of the deadly surges that occur when lava breaks loose and hot gases that had been trapped are released. The heat of the gases helps convert the avalanches into mobile pyroclastic flows—clouds of gas, pumice, and ash that can travel at 80 miles an hour and contain heat of 800°C. (One such cloud swept down from Mount Pelée on the Caribbean island of Martinique in 1902 and wiped out 30,000 people in the city of St. Pierre.)

As the three walked in the valley, a portion of the lava dome at the top of the mountain collapsed with a sound like thunder. Instead of a small pyroclastic flow cascading down a ravine, a roiling cloud of intense heat filled the entire valley. The Kraffts, Harry Glicken, 40 journalists, taxi drivers, firemen, and farmers farther down the mountain were all killed.

The day I arrived, a Japanese newspaperman and I drove near the valley to see the seared vegetation and burned houses. As we snapped photographs, more lava broke loose and rolled down the mountain in another huge pyroclastic flow, 200 feet high, perhaps a quarter of a mile wide, its leading wall of searing heat boiling upward from the bottom—the most terrifying force of rampaging nature I have ever seen. We sped away in the newspaperman's van, but it was easy to see that, for those directly in the valley days before, there had been no possible escape. The flow I saw destroyed 73 more houses.

"The Kraffts and Harry Glicken knew the danger of being there, but they were dedicated to learning more about volcanoes," said Bob Tilling, a USGS volcanologist at Menlo Park, California. The Kraffts' parents donated to the volcanologic community the rights to the Unzen footage, for use in more safety films.

More than a decade before Unzen's latest blast, a volcano in the United States had popped a tragic surprise. In 1980 Mount St. Helens, a hundred miles south of Seattle, showed all the classic signs of an approaching eruption as magma moved up—frequent small earthquakes and minor puffs of steam escaping from the top of the peak.

Roger Ressmeyer/CORBIS

Deadly Beauty

Princely Cotopaxi in Ecuador, one of the highest active volcanoes in the world, stands 19,460 feet high. The heat from even a small eruption could suddenly melt vast quantities of ice and generate torrents of mud called lahars. In the past, lahars have flooded surrounding lowlands up to 200 miles away.

Then the unexpected happened. An earthquake caused the north side of the mountain to slide away, sending tons of rock, earth, and ice 15 miles down the North Fork of the Toutle River Valley. Removal of the slope opened a fissure in the volcano, and a lateral pyroclastic explosion blew out the northern flank of the mountain, leveling trees as far as 17 miles away. The violent blast and the volcanic avalanche killed 57 people.

The collapse of Mount St. Helens prompted volcanologists to take a closer look at the terrain around other volcanoes and to realize that these scenic cones are less stable than they appear. In Colima and Jalisco states in Mexico, volcanologist Ana Lillian Martin-Del Pozzo showed me a series of abrupt hills jutting hundreds of feet above a level plain.

Seven miles away towered Nevado de Colima, a sprawling mountain with a peaked volcano smoldering on one rim and occasionally spitting thick lava that dribbled down the growing cone. When the volcano was much larger, thousands of years earlier, one side had collapsed. The hills we stood before were the remnants of the cone that had given way and slid onto the plain. Sprinkled with fine ash, the hills had grown trees and grass, giving them the look of giant bread loaves covered with green mold.

"The volcano continues to rebuild," said Martin-Del Pozzo, "and the cycles of lava flow, followed by fine ash, followed by more lava, will go on until the mountain can no longer support itself, and it will collapse again." Some material from an earlier collapse appeared to have reached the sea, 55 miles away. The city of Colima, population 200,000, lies less than 20 miles from the mountain.

The collapse of Colima pales by comparison with the havoc in the Hawaiian Islands. In the 1960s, USGS

Agenzia Contrasto/SABA (both)

Too Close for Comfort

A river of fire pours down the slopes of Mount Etna toward the Sicilian village of Zafferana (above). The opening of a vent in 1991 threatened to pour lava into the village, which had suffered volcanic damage in previous centuries. Italian troops placed explosives (opposite) to divert the flow. United States Marines dropped concrete dams from helicopters. Finally, when the lava was diverted into an excavated channel and the natural channel was blocked, the village was saved. Etna stopped coughing lava after 473 days.

volcanologist Jim Moore was studying underwater maps made from echo soundings when he noticed huge blocks lining the ocean floor for miles around the islands. He later verified them as parts of the land that had collapsed earlier and slid into the sea. Some of the blocks had tumbled 125 miles and were as big as Manhattan Island.

The verdant vacation meccas visited each year by millions of tourists are only the latest in a string of islands that reach nearly to Kamchatka. They grew from a hot spot, a blowtorch section of magma that keeps boring up through the crustal plate and building islands. The hot spot remains in the same place, but since the plate is moving northwest at a rate of four inches a year, it has left a trail of islands whose remnants are discernible on the ocean floor. The earlier ones eventually collapsed and wore away. Major breakdowns that threaten today's islands occur perhaps every 100,000 years, says Moore, but on the south side of Hawaii, a minor slump is due. Like sand settling in an hourglass, the growing mass of some 500 acres that Kilauea has added to the island over the past decade will eventually cause adjustments.

"With the Global Positioning System (GPS) we now look at a certain location on the earth at various times from a satellite and see if it has moved," said Moore. "A slab about 31 miles long on the south side of Hawaii is sliding toward the sea at rates up to almost 4 inches a year. Eventually it will slump dramatically."

"In many ways we humans have been fortunate where volcanoes are concerned," says Bob Tilling. That good fortune is in jeopardy because of the human population explosion and potential volcanic blasts larger than any seen in history.

"There are, on average, 50 volcanoes erupting above *(Continued on page 76)*

Paul Chesley (above and opposite); Roger Ressmeyer/CORBIS (below); Bill Nation/SYGMA (following pages)

The Earth Moves

The dome of lava named Panum Crater—part of Mono Craters—rose out of the desert 600 years ago (left). Just to its south, in what is now Long Valley, California, an eruption 760,000 years ago scattered ash halfway across the continent and left a collapsed 10-by-18-mile-wide caldera. A scientist (opposite, lower) monitors the restless ground for signs of another eruption in an area now dotted with ranches, towns, and a ski resort. Laser beams measuring movements of the resurgent dome indicate that more magma is moving into the cauldron underneath. A hydrologist checks the temperature and flow rate of hot springs in the valley (below), also indicators of subsurface magma. To prevent misunderstandings and panic, scientists keep local residents informed. Just as volcanic eruptions have vastly altered local landscapes in the past, geologists don't doubt that such eruptions will happen again, but possibly not for hundreds of thousands of years.

On broken Antelope Valley Freeway lies the body of a Los Angeles police officer who rode his motorcycle off a collapsed overpass after California's Northridge earthquake of 1994.

Rick Rickman/MATRIX; Les Stone/SYGMA (upper); Eugene Fisher (opposite)

At the Mercy of Unstable Terrain

The junction of major tectonic plates along the San Andreas Fault and a complex network of branching faults keep life uncertain in California, where earthquakes have brought death and massive damages in recent years. Rock slippage on an unknown fault buried beneath Northridge in Los Angeles caused 60 fatalities on January 17, 1994, and created a surface rupture on a hillside (opposite). Broken city gas and water lines resulted in geysers of flame and fluid (opposite, lower). Collapsing apartment building overhangs mashed cars (below).

Andrea Booher/FEMA (both)

sea level every year, and this rate has been pretty constant for centuries, so there's no reason to believe it is going to stop," Tilling said. "But we've had rapid growth in human population over the past hundred years. Now millions of people live on the flanks of volcanoes that merely doze at the moment, such as Mount Vesuvius just outside Naples, Italy. And eruptions from Vesuvius have been small, compared with some."

When does one suggest that people leave? Dave Hill knows something about the dangers of prediction. In the early 1980s he and other geologists noted with concern that the ground of Long Valley caldera in east-central California was rising and that fumeroles were spitting gases. The area was shaken by swarms of minor earthquakes and two quite large ones. Several hundred millennia before anyone studied volcanoes, Long Valley

High Cost of Faulting

Damages after the Northridge spasm exceeded 40 billion dollars, a record for U.S. earthquakes. The magnitude 6.7 temblor destroyed more than 3,000 homes, toppled 10 highway bridges, closed 3 major freeways, and demolished part of a shopping mall. It defaced the Kaiser Permanente building (above) by shaking loose its facade and damaged the interior and furnishings, as well. Bricks falling from trembling buildings also smashed vehicles; an unwelcome load crunched a van (opposite). Geologists predict an even bigger quake in the future, with damages of staggering dimensions.

caldera was formed in one of North America's largest blasts, scattering ash as far east as Nebraska and creating an oval-shaped depression 10-by-18 miles wide. Could such an eruption be coming again, to an area now well populated with ranches, towns, and a ski resort?

"We wrote a report suggesting that there was sufficient concern that local officials in the valley should be contacted," Hill told me at his office in Menlo Park. "It recommended that the activity be watched closely and added that an explosion was probably not imminent. But a newspaper reporter got hold of it, and, before the local officials heard anything from us, people in the area had already read in the *Los Angeles Times* that we were about to warn them of the possibility of an eruption. Understandably, they were very upset. Business people worried that clients would avoid the valley. We heard arguments like, 'I've lived here five years and nothing has happened'.

"We realized that we had to improve our communications. And we needed to remember that we were talking with laymen not used to thinking of earth changes in terms of thousands of years. So now we give regular seismic reports on the area in the local papers."

What is happening now in Long Valley caldera? Hill wheeled around in his chair and tapped a few keys on a computer. An outline of the former caldera appeared, with certain sections shaded in green. "The green areas show where the most seismic activity has occurred in the past three days," he said. "Let's see how bad they were."

He tapped a few more keys, and figures replaced the caldera. "Seventy-two earthquakes in the past three days, most of them small, although this one registering three was probably felt by some people. When the frequency hits twenty earthquakes in an hour, an alarm goes off and we start watching it more closely. The uplift has continued—the area has risen almost 24 inches since the early 1980s—but we're not currently in any kind of warning phase."

Consider the effects of another eruption

After the Quake, Shocks of Grief

The anguish of loss brings Hyun Sook Lee's hands to her face as a fireman tells her that her 14-year-old son has died in the Northridge quake. Her husband was also killed when their apartment collapsed. Of the 60 people who lost their lives in the quake, 16 died when the top two floors of the three-story Northridge Meadows apartment complex fell onto the first floor. Along with the tragedies came tales of luck: Sleep saved a man who dozed off watching television in his living room; debris crushed his bed. Insomnia saved another, a man who was reading in the kitchen when an eight-foot-tall bookcase fell onto his bed.

of the scale that formed Long Valley caldera or Yellowstone caldera. Pinatubo spat about 1.2 cubic miles of magma, disrupting thousands of lives and temporarily altering the climate. Long Valley caldera vented 144 cubic miles of material 760,000 years ago, and Yellowstone caldera vented some 240 cubic miles of ash 600,000 years ago. Another eruption of this scale would truly have a global impact.

"If that much magma moved close to the surface, we would be able to detect it and see that it was coming," Hill said. "But you really can't worry about an eruption of that size. We're talking about geologic time. We probably won't see one, nor will our children or grandchildren."

But will it happen? He laughed, a geologist's amusement at a layman's incredulity. "Oh, sure, it will happen again someday. It always has."

Les Stone/SYGMA

Unlike volcanoes, a large earthquake gives little warning of its approach. The plates that meet do not slide past each other smoothly. "Think of hooking a spring to a brick, and then using the spring to pull the brick across a tabletop," a geologist told me. "Because of its weight and its friction against the table, it would slide, then stop, slide, stop. It's similar with tectonic plates because of the pressure of one against another, and their slow movement—slide, stop, slide, stop."

When Californians talk about "waiting for the big one," they mean an 8.2 earthquake or larger. It will happen when the North American and Pacific plates, continent-size blocks now hung up on each other at sections along the San Andreas fault system, jerk loose. In recent years they have had two disastrous preludes, the quakes centered at the peak called Loma Prieta near San Francisco in October 1989 and at the community of Northridge in Los Angeles in January 1994.

"Both Loma Prieta and Northridge are points where the San Andreas makes a bend toward the west," said Jerry Eaton, USGS geologist at Menlo Park. "These bends create a zone of compression as the plates try to slip by them. Both earthquakes were a result of movement along the San Andreas fault system, but the slippage occurred on smaller, auxiliary faults pressured by movement along the main fault zone. Unfortunately, the release of these smaller faults may mean that there is now less resistance to a bigger slip on the main fault."

Scientists can now track the movement of the two plates before an earthquake occurs, thanks to the same GPS by which Jim Moore watches Hawaii's south slab creep toward the sea. "We look for surface deformation," said Eaton. "There are places along the San Andreas where you

Fear: Slow to Subside

Faith in their homes shaken, many Northridge survivors camped outside (below), fearing earthquake aftershocks. Geophysicist Andrea Donnellan, of NASA's Jet Propulsion Laboratory, checks instruments at a Global Positioning System (GPS) site (opposite) that confirm survivors' fears. Satellite measurements showed that faulting pushed the ground up nearly 15 inches and moved it more than 8 inches northwest. Fear of price-gouging by opportunists became a side effect. A note posted (opposite, lower) declares fair rates for those needing to store belongings.

Eugene Fisher (above and right lower); Roger Ressmeyer/CORBIS (right upper)

could put one of your feet on the North American plate and one on the Pacific plate; we know the location of the boundary in such places that precisely. You'd have to stand there for months to notice that your feet had moved, but GPS shows that movement is actually taking place in that spot. Near Loma Prieta what we saw was not slippage along the San Andreas but warpage of rock on both sides of it.

"So we can see compression building up; we just don't know when it is going to let go, and exactly where. What we hope is that by watching the process carefully in a number of places, we can eventually see a pattern. As we develop better tools for measuring movement, we may be able to improve our ability to predict."

How does an earthquake specialist feel when an earthquake happens? Jerry Eaton's house is sturdy and built on solid rock some distance from the epicenter of Loma Prieta. It remained standing, but like millions of San Franciscans he rode through a severe shaking, with thoughts a bit more clinical than most people's.

"I was excited," he told me. "I'm sure my pulse went up. I had a sense of things bending in my house, but I don't know if they actually were—or if the rapid movement just made them seem that way, like this...," he said, waggling a pencil rapidly between his fingers. "I remember trees outside whipping back and forth. Mostly I just sat there and thought about what I was experiencing."

On the other hand, fireman Tim Petersen didn't have time to think about what was happening. Driving his pickup truck on the Nimitz Freeway in Oakland, he didn't see the earthquake coming. His section of the freeway fell in the first five seconds, and his truck ended up just 18 inches high. He lay trapped for more than five hours with six broken bones and a collapsed lung before other firemen dug him free. "I look at things differently now," Tim said later. "It can end so fast."

The Loma Prieta quake killed 62 people, while Northridge took the lives of 60. Damages differed considerably. Although Loma Prieta was the larger quake at magnitude 7.0 compared with 6.7 at Northridge, its damages totaled some 10 billion dollars, compared to more than 40 billion dollars in the Los Angeles area. For measuring earthquakes, scientists now prefer the moment-magnitude scale over the Richter. The moment magnitude is a measurement of total energy released by earthquakes, calculated in part by multiplying the area of the fault's rupture surface by the distance the earth moves along the fault.

The Northridge fault broke directly beneath a heavily populated commercial area that was built on weak sedimentary rock. It shook more violently during the quake than did the firmer rock near the Loma Prieta epicenter. A thousand buildings were demolished by Loma Prieta, and a major freeway collapsed. Northridge destroyed or rendered uninhabitable more than three thousand homes besides toppling ten highway bridges and closing three major freeways. And neither of these was "the big one," expected in California. The biggest one so far in North America hit Alaska in 1964.

The northward-moving Pacific plate dives beneath the North American plate along the arc of the Aleutian Islands and southern Alaska. The Good Friday quake of March 27, 1964, occurred in Alaska when the Pacific plate fractured with one of the biggest jolts ever recorded anywhere, an earthquake with a magnitude of 9.2. Only a Chilean earthquake of 1960 was bigger, at 9.5.

Earthquakes had long been suspected of parenting giant waves known as tsunamis, but this connection was nailed down as fact after the Alaska event in 1964. "The faulting caused an uplift of some 38 feet on the ocean floor," said USGS geologist George Plafker, sent to Alaska after the event. "That caused a tsunami that reached Anchorage, but the city is 100 feet above sea level and the wave caused little damage. We knew the wave arrival time at coastal sites and the depth of the ocean, so we were able to calculate the speed of the wave and to determine exactly where it originated. When we did that, it was right on the 300-mile-long offshore zone where the greatest uplift happened in the quake."

Volcanoes, too, can cause tsunamis. It was not the infamous explosion of Krakatau in 1883—heard 3,000 miles away—that caused most fatalities, but the tsunami it created in the Sunda Strait between Java and Sumatra, which drowned 36,000 people. Two hundred years before part of Unzen's new dome collapsed in 1991, one whole side of the mountain came down, covered the town of Shimabara, and slid into the sea, resulting in a powerful tsunami that hit many populated coastal areas. The combination of Unzen's partial collapse and the tsunami killed 15,000 people; it still holds the record as Japan's worst volcanic disaster.

While out to sea, the energy of tsunamis is spread out in the depths, and the wave may pass under a sailboat with little more disturbance than a normal swell. The greatest height and breaking crest happens when it reaches coastal shallows—hence the possibility of a surprise attack thousands of miles away.

The Pacific Tsunami Warning Center was set up in Hawaii after a 1946 earthquake in Alaska sent a big wave to Hawaii that killed 159 people and caused extensive damage. Now the center, operated by the National Oceanic and Atmospheric Administration (NOAA) receives seismic and tide-level information from stations all around the Pacific Rim and within an hour can warn coastal dwellers if a tsunami is moving across the Pacific. A second warning center has been established in Alaska to send regional alerts to people living along the coasts of Alaska, Washington State, Oregon, and California. The U.S., Japan, Russia, French Polynesia, and Chile are the only nations that now have regional systems that can warn people within 12 minutes. But the system is not foolproof.

"Anyone within 10 minutes of where a tsunami originates is still at risk," said Mike Blackford, director of the Tsunami Warning Center in Honolulu. "If it takes five minutes to get that warning out, that leaves people only five minutes to get to

Carving Landscapes

The slow, steady grind of one massive plate against another is traced by California's Wallace Creek, named for a geologist who first studied its detour along the San Andreas Fault. Where the North American and Pacific plates meet on the Carrizo Plain, the northwestward march of the Pacific Plate at the top of the photograph has moved the creek more than 426 feet to one side. The rerouting, geologists have discovered, has taken 3,700 years, in roughly 33-foot jumps, during a dozen earthquakes.

Georg Gerster

20is10x
KOFY20
Grant Av

Oakland Freeway Exacts Heavy Toll

Not since San Francisco's devastating earthquake of 1906 and the San Fernando quake of 1971 had a tremor brought such horror to California. During the Loma Prieta quake of October 17, 1989, segments of Oakland's Nimitz Freeway (opposite) pancaked onto lower traffic decks, crushing vehicles and their occupants. A car flattened to its axles (below) shows the force of tons of concrete. In an instant, 42 people died, out of 62 killed Bay-wide by the magnitude 7 quake. A slab collapsed just short of one driver's tractor-trailer (above). Analysts blamed the collapse on inadequate connections between columns supporting the upper and lower roadways. Loose soils also amplified the motion of the vibration.

Monte Farnes; James A. Sugar/BLACK STAR (upper); THE PHOTOFILE/Dave Bartruff (opposite)

high ground."

NOAA oceanographer Eddie Bernard has devised a system for local warnings, by which seismic information is telemetered to a computer. "If the vibrations exceed a certain magnitude, a message automatically goes to a satellite," he explained. "An alert travels from there to a number of receiving stations around the region and alarms are sounded along the coast. It cuts the warning gap down to two minutes after a tsunami might have started." The system is now being used experimentally in Chile, where the subducting Pacific plate causes numerous offshore earthquakes.

P.F. Bentley/BLACK STAR ; James A. Sugar/BLACK STAR (upper); Dan Lamont/MATRIX (opposite)

Any violent disturbance to the fragile crust on which we live can strike with a suddenness that defeats the best of warnings. When a 45-second earthquake shook the Peruvian town of Yungay in midafternoon of May 31, 1970, the sheer west face of the peak called Nevado Huascarán avalanched off the mountain. Fluidized by ice and snow, the mountainside traveled the eight miles to the town in about three minutes. The town of Yungay was almost completely covered, sometimes to a depth of 30 feet, and 18,000 died there.

You may never have heard of Yungay. Are we sometimes more fascinated by what earth's forces can do to modern, man-made edifices than by the horrible toll they often take on human life? Both the amount of property destroyed and the number of lives snuffed out by natural forces may mount as the world population continues to climb. Greater numbers of people living on earth's crust may mean that space becomes more precious and property values increase. Meanwhile, we cluster in ever larger numbers and build more structures on precarious terrain.

"National disasters are beginning to affect our economies in a very real way," said George Plafker, the tsunami special-

On Shaky Ground

Unstable soil collapsed this house (above), one of many wooden structures built over man-made fill in San Francisco's popular Marina district. Liquefying in the Loma Prieta quake, sandy soils saturated with water became quicksand and allowed unreinforced buildings to settle and break. Violent shaking also caused an upper section of the San Francisco–Oakland Bay Bridge to fall (opposite, lower), killing one driver. Had bolts not sheared (opposite, upper), swaying trusses might have pulled down even more of the vital span.

ist. "More than 10 billion dollars for Loma Prieta, more than 40 billion dollars at Northridge. What will the costs be when a really big earthquake hits a highly populated area?"

We had that conversation days before the Great Hanshin Earthquake struck Kobe, exactly a year to the day after the Northridge quake. The Hanshin quake was like the Northridge quake in that it occurred along a so-called minor fault. Area officials had dismissed it as a source of trouble because it had been fairly quiet for half a century. But in Kobe the human fatalities and damages reached the proportions Plafker had feared. Japan, with its efficient, industrialized economy and favorable balance of trade, is one of the world's richest nations; yet, its budget and national psyche were challenged by the loss of 5,500 citizens *(Continued on page 93)*

Nightmarish Waves Spawned by Violent Earth

Offspring of earthquakes and volcanoes are the huge waves known as tsunamis. Thousands drowned when the infamous blast of Krakatau in 1883 sent giant waves hurtling across Indonesia's Sunda Strait. This 20-foot wave (opposite) caused little damage when it churned over the same waters nearly a century later, after a smaller eruption and pyroclastic flows from Anak Krakatau. Faulting on the ocean floor can generate great waves that move unnoticed before rising up near land. Warning systems can now alert coastal residents hours ahead of sea-spanning tsunamis, but an earthquake off Hokkaido, Japan, in July 1993, gave nearby villagers only minutes to escape. A tsunami hit the tiny island of Okushiri, demolished houses, tossed boats ashore (below), and killed more than 200 people.

Tetsuji Asano/ASAHI SHIMBUN; Dieter & Mary Plage (opposite)

Coastal Peace Can Be Quickly Shattered

On a clear, calm day the Humboldt Bay power plant and this spit of land in front (opposite) near Eureka, California, seem situated in placid surroundings. But the junction of three tectonic plates make the Eureka area subject to earthquakes and tsunamis. Just offshore, the Gorda plate dives under the North American plate at the same time it is butted from the south by the northward-moving Pacific plate. At the western end of the Pacific plate, Japan has seen the effects of one plate pushing under another. The 1993 earthquake off Hokkaido shook houses onshore minutes before a large tsunami came rolling in. Firestorms started by broken gas lines in the town of Aonae (below) leveled about half of its 680 homes.

Roger Ressmeyer/CORBIS; Tetsuji Asano/ASAHI SHIMBUN (lower)

Paul Chesley (all)

Quake Rehearsals

Knowing that earthquakes are inevitable on their tectonic crossroads, residents of Tokyo rehearse their preparedness on Disaster Prevention Day, September 1. A girl practices mouth-to-mouth resuscitation on a doll representing a baby overcome by smoke (opposite), while others learn how to use fire extinguishers on real flames (opposite, lower). Students try on clear plastic bags (below) containing just enough oxygen for the crucial seconds needed to escape a smoke-filled room. Japanese memory hearkens back to the Great Kanto Earthquake of 1923. Although shaking brought down buildings in Tokyo and Yokohama, most death and damage came when fires from hibachis spilled hot coals in wooden houses. A firestorm killed 30,000 crowded into one park. The final estimate: 143,000 dead or missing.

and costs estimated at 200 billion dollars.

The city of one and a half million people, while best known as the origin of tender, beer-fed beef, was an important producer of steel and electronic products. Twelve percent of all Japan's exports left via Kobe's ports, which handled 10,000 oceangoing vessels a year. Many port facilities were knocked out by the quake.

In the city itself, structural damage was not total, except where fires obliterated whole blocks. Elsewhere, buildings seemingly untouched sat next to others fractured, windowless, and vacant as skulls. The difference lay in construction, and the capriciousness of earth's vibration.

"Kobe had a large number of concrete buildings that lacked the steel confinement—consisting of hoops or ties—that holds concrete together under seismic forces," said Craig Comartin, a structural engineer from San Francisco, who visited Kobe the day after the quake. "Building codes didn't require confinement until the 1980s, and the structures

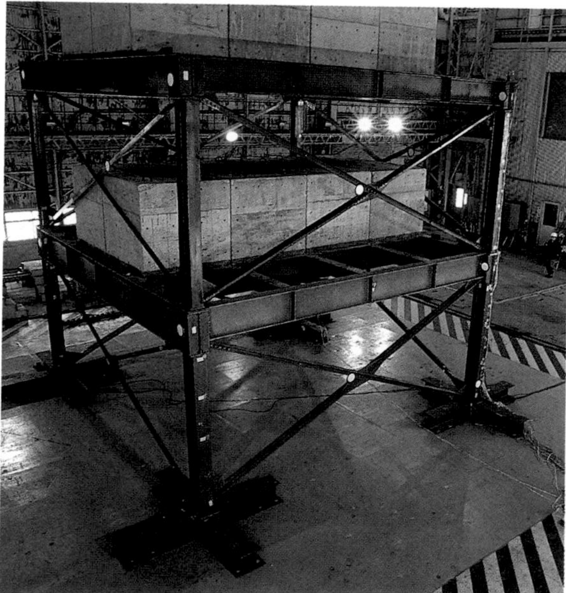

Paul Chesley; Michael Yamashita (upper)

Baring Weaknesses

A building canted toward the street by the Kobe earthquake in 1995 (above) was an older steel building not up to modern standards. Its strength proved insufficient to withstand seismic forces. Japanese engineers have been developing modern design standards such as the braced steel frame being tested on a shaking table (left) at Japan's National Research Institute for Earth Science and Disaster Prevention at Tsukuba. A researcher examines damage to a concrete column heavily weighted and subjected to quakelike vibrations (opposite, lower). Collapse of road-bearing columns (opposite, upper) toppled the Hanshin Expressway in the Kobe quake.

that had it performed pretty well. Without it, concrete tends to explode when subjected to earthquake forces.

"That's what happened to the columns holding up the freeways there and in California. Now they're retrofitting the columns in both places with steel confinement to make them stronger."

Some structures in earthquake country are being fitted with dampers—alternating layers of steel plates and rubberlike polymers that allow the footings of bridges and buildings to slip but not break and the structures to shimmy and sway, perhaps harmlessly.

"Damping shows real promise," said Comartin, "as long as the building doesn't bang into anything else as it's moving back and forth. Buildings perform better during an earthquake if you make them symmetrical, something architects seldom like to do.

"A very large tremor occurring right under a structure built for earthquakes can still damage it. No building is earthquake-proof, but you can lower the risks."

I remembered the anguish of a Kobe landlord, Toshu Kiyomura, who thought his apartment building was safe from temblors. I saw him praying, head bowed, as rescue workers removed the last of five bodies from his toppled five-story building. Two tenants had been removed earlier and died in the hospital. "This wasn't supposed to happen," he told me, sadness lining his face. "The foundation was anchored deep in the earth."

"Our family lived on the top floor," he said, reliving the event. "Suddenly the building began jumping up and down and twisting at the same time, like a dancer. Then it fell over and I became unconscious. I heard my wife calling my name and for a moment thought it was just a bad dream. Then I opened my eyes and saw the wreckage and crawled out a

Paul Chesley; Haruyoshi Yamaguchi/SYGMA (upper)

The High Price of a Living Earth

A volunteer fireman mourns the loss of his best friend, a baker and fellow volunteer, who died in the Nagata Ward in Kobe in 1995. This section of the city was leveled by fire after being shattered by the quake. Firemen hampered by broken water mains had to stand by as flames burned houses with people trapped inside. Of all earth's cataclysms, few rival the power of earthquakes to disturb our equilibrium.

Michael Yamashita

window onto the street. Out of 35 people living here, 7 died."

The living needed physical as well as spiritual help, and friends and relatives responded. Because of badly congested motorized traffic, streams of bicycles, motorcycles, and pedestrians poured in from outlying areas with food, liquids, clothing, and blankets. Soup lines sprang up, merchants sometimes donating the food. I saw shops gaping open from the quake with merchandise unprotected—clothing, appliances, liquor—but looting was virtually unknown.

Five days after the quake, electricity and gas were still off, and water was still being hauled in. Thousands of people were camping in schools. I walked along a deserted dock, leaping across gaping cracks in concrete wharves, hearing the sad slap of wavelets against idled ships.

A half-submerged car clung to a parking spot tipped partway into the bay. Behind me were piles of rubble, tilted buildings with broken windows like empty eye sockets, toppled freeways. A trembling earth had brought a prosperous, modern city to its knees.

I have seen destruction from wartime bombs and damage from earthquakes and volcanoes. Somehow a destroyer that leaps from inner earth is more shocking, perhaps because it cannot be stopped and its origins cannot be seen. Geologists do not see an enemy.

"Scary and costly as they are, we have to be grateful for eruptions and plate movements," said seismologist Jerry Eaton at Menlo Park in California. "They built the ground we live on, and they continue to build it. If the earth ever cools, this ground building will stop, and with time everything will erode, become level, and a thin sea will cover it all.

"Earthquakes and volcanoes mean our planet is alive."

From a sky dark with menace, a tornado stings a north Texas prairie.

by Thomas Y. Canby

For three days the massive storm Alberto has dumped record rainfall onto Georgia farmland draining into the Flint River. Now the bloated Flint, 20 feet above flood level and rising, is leaping its banks inside the city of Albany and surging through a community called the Bottom.

A flood warning has gone out in Albany, but not everyone has made it to safety in time. Not the five desperate souls trapped in the tiny frame bungalow at 1501 South Jefferson Street: grandmother Eloise Barnes and four of her family, including two children, along with the cat, the chow, and four pups.

"We couldn't get out—the water outside was rising so fast we couldn't push the door open," Mrs. Barnes recalls of that nightmarish Thursday in July. "Water gushed in around the door and began filling the house. The others were frightened; I just prayed. Thirty or forty minutes went by, and the water got up to our necks."

Her daughter Dorothy, mother of the children, remembers the flood's foulness from agricultural and other wastes: "The smell of the water, the horrible taste—you thought you'd die."

"Then the door opened," continues Mrs. Barnes. "I don't know how—the Lord opened it. A bulldozer was outside—it had been moving up the street helping people. We scrambled into the scoop. But we'd forgotten the dogs and cat. My son-in-law swam back through a window and brought them out.

"My son had climbed an oak tree

Najlah Feanny/SABA; Chris Johns (opposite); Warren Faidley/WEATHERSTOCK (preceding pages); NASA (following pages)

across the street, and we got him into the bucket. Three kids were hanging from a light pole. We rescued them.

"We lost everything. But that dozer saved our lives."

Thirty persons lost their lives—many in flash floods—that July 1994 when tropical depression Alberto stalled over Georgia, Alabama, and Florida. As much as 21 inches of rain fell in a day. Tens of thousands of people were forced from their homes. Property damage climbed to half a billion dollars.

Flash floods claim nearly 140 lives a year to stand as the number one weather killer; indeed, they take the highest toll of all the weather-related disasters. Next most lethal is lightning, with an annual average of 93 deaths and 300 injuries. Tornadoes kill about 80 Americans a year and inflict 1,500 injuries. Hurricanes take a similar toll. *(Continued on page 106)*

Storms' Deep Scars

Sudden and severe storms exact a costly toll psychologically as well as physically. Catholic sisters on Kauai (above) comfort bewildered students amid the wreckage of their school, leveled when Hurricane Iniki slammed into the Hawaiian island in 1992, with winds of 130 miles an hour. Almost two years later at Americus, Georgia, tropical storm Alberto caused Town Creek to leap its banks and wash out a section of North Mayo Street (opposite). Alberto's damage was estimated at half a billion dollars; Iniki's, almost four times that.

Ominous eye of Hurricane Emilia peers from coils of cloud whose winds reached 160 miles an hour, making it one of the most powerful hurricanes recorded in the central Pacific. Astronauts of the shuttle *Columbia* monitored the monster as it prowled harmlessly south of Hawaii.

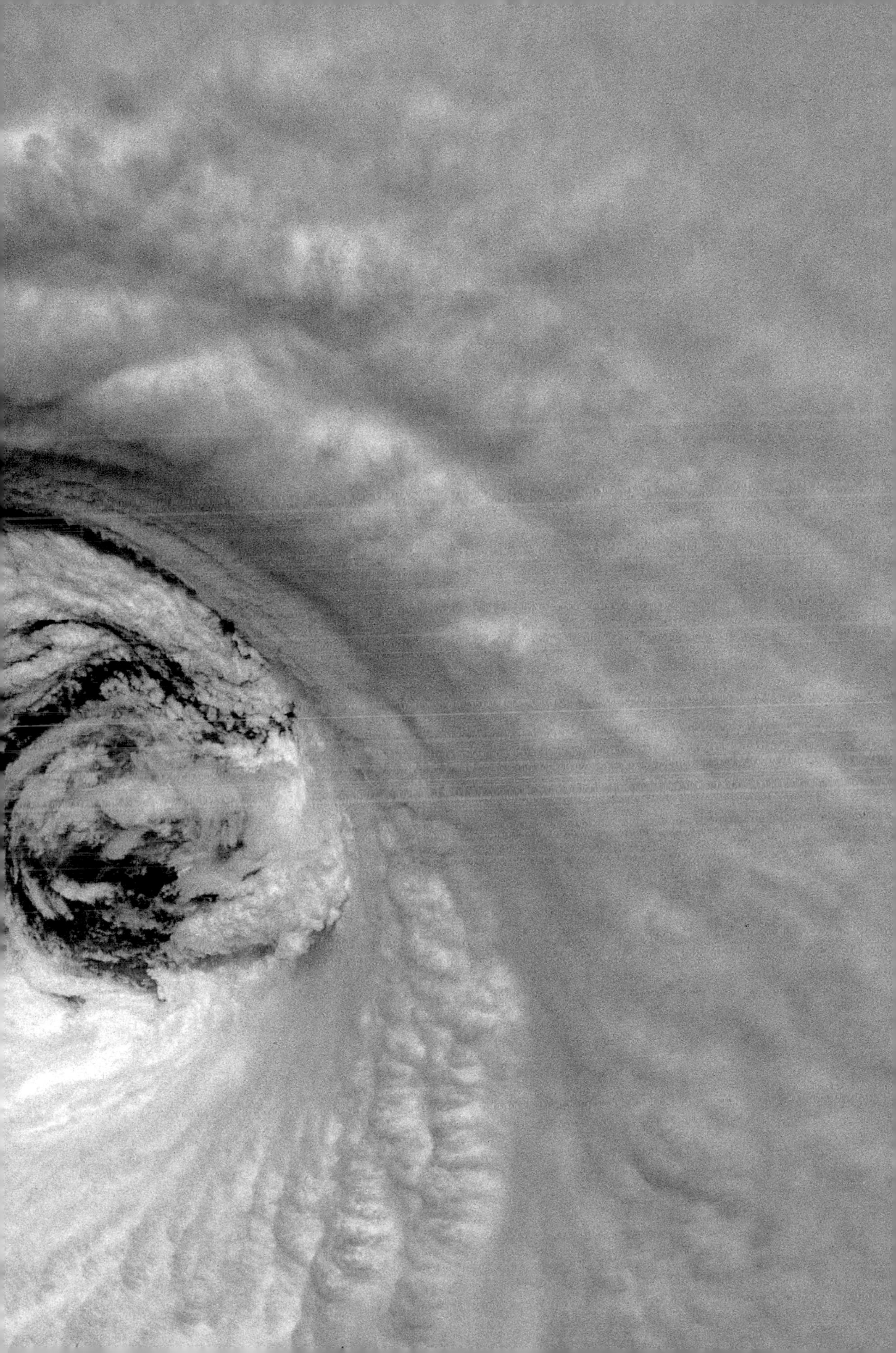

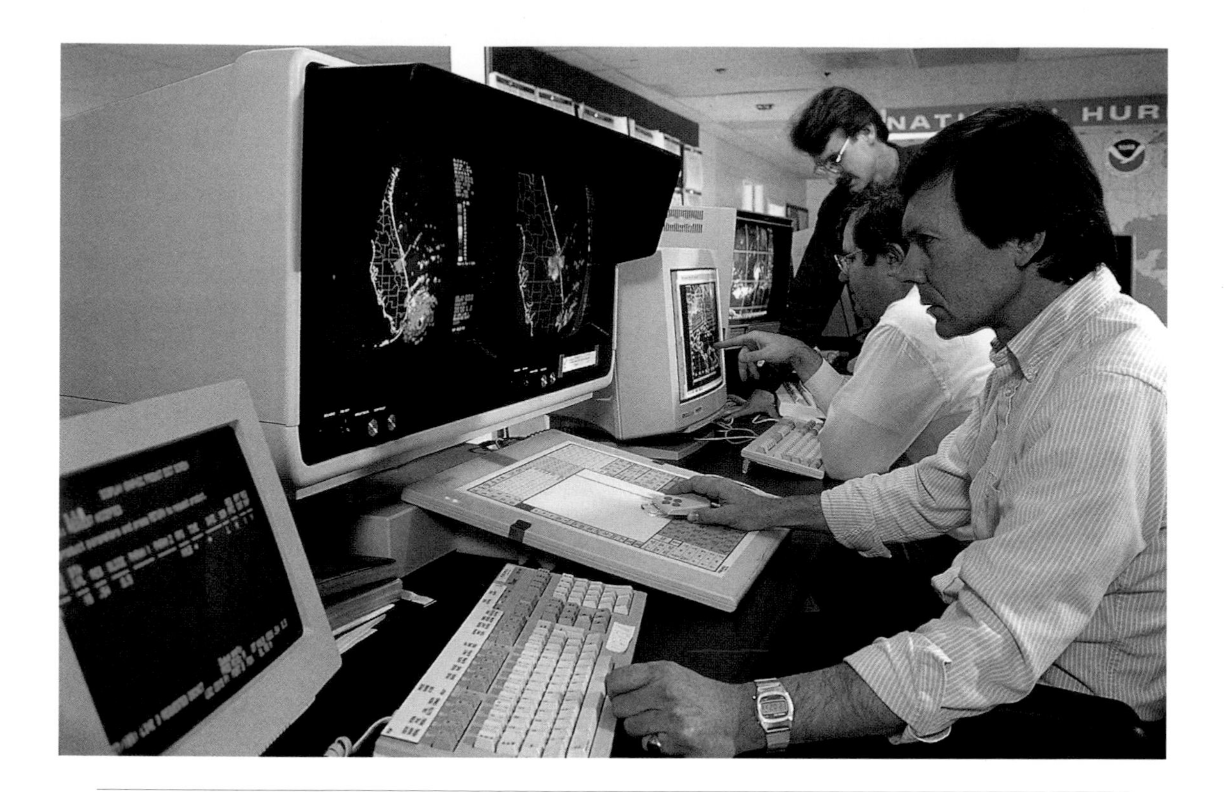

Always on the Alert—Analyzing Data

Landfall of Andrew—the hurricane that mangled South Florida in 1992—fills the computer monitor of meteorologist Colin McAdie (above) at the National Hurricane Center, now located in Miami. Observation of the future Andrew had begun ten days earlier over West Africa, before it moved into the Atlantic and became a tropical storm like Ernesto, shown in a 1994 satellite image (below). Aided by radar, aerial observation, and satellites, hurricane specialists can track storms with precision, but accurately predicting time and place is difficult. Warning of Andrew 21 hours before landfall facilitated the evacuation of the Florida Keys (opposite).

Mike Clemmer (both); Lannis Waters/*PALM BEACH POST* (opposite)

Hurricane Andrew left a 32-billion-dollar repair bill in South Florida and Louisiana in 1992. Hail, destructive to property, seldom kills, but it can: An 1888 hailstorm pummeled parts of India, taking 246 lives. In wintertime, fierce snow and ice storms paralyze vast areas. The awesome powers of severe storms and the immense losses of property and lives that result from these raging forces of nature give mere humans pause for thought as well as a lot of work to do—in both preparing for and cleaning up after them.

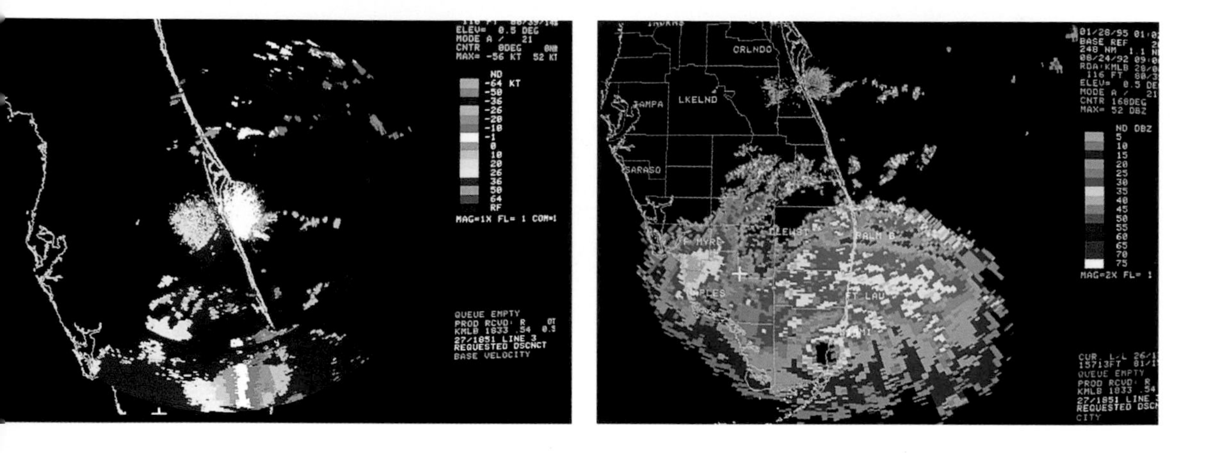

The United States suffers more severe storms and flooding than any other country on earth. NOAA calculates that 10,000 severe thunderstorms and 1,000 tornadoes strike each year. In addition Americans contend with 5,000 floods and withstand the buffeting of, on average, two hurricanes annually.

Why this overly generous allotment of inclement weather? The large size of the United States and the fact that it straddles a whole continent are, of course, important factors. Another factor is North America's peculiar geography. Other continents, notably Europe and Asia, contain major east-west mountain chains that serve as barriers against interactions of cold polar fronts and hot, moist air from the south.

In North America the central Plains sweep virtually unbroken from polar north to sultry south, inviting explosive collisions of vast air masses. These generate many major thunderstorms, tornadoes, and violent winter storms.

Another vulnerability traces to the nation's immense frontage on the Atlantic Ocean and the Gulf of Mexico. This superb sweep of popular beachfront, home to ever more millions of us, poses an almost unmissable target for storms that gather strength in the tropical Atlantic and smash ashore in North America as an Andrew or an Alberto.

To reduce the damage wrought by these storms, NOAA and its major weather arm, the National Weather Service (NWS), focus their efforts on neutralizing a storm's most dangerous weapon: surprise.

Through storm prediction, or forecasting, the NWS can now give two to three days' advance notice of the possibility of a hurricane strike and 12 to 24 hours' notice of the probability of tornado development or of conditions that can create wind shear. Precise warnings, however, are generally more limited. For hurricanes such warnings are provided about 24 hours in advance of the event, while tornado and wind-shear warnings are frequently limited to a few precious minutes.

New Doppler radar systems are expected to increase warning time for larger tornadoes to 15 to 20 minutes

When Winds Hit

Darkness gathers, debris flies, and trees bow as Andrew lashes Fort Lauderdale Beach (below). Winds, once invisible to radar, can now be picked up by NEXRAD, one of the NWS's first operational Doppler radar systems, in Melbourne, Florida. An image (opposite, left) shows the northern fringes of wind circulation around Andrew. Another radar view (opposite, right) reveals the intensity of rainfall in the hurricane as the storm's eye comes into Biscayne Bay. Yellow shows heaviest rain.

before they strike a particular community. Even with the advancing technology, weather systems are so complex that they still can defy the largest supercomputers and the most experienced meteorologists. Yet, steady and sometimes dramatic improvements have occurred, enabling more accurate forecasting of where and when killer weather will strike.

When the great rainstorm Alberto struck the Southeast, its toll of deaths followed a familiar pattern. A majority of those who died were lost to flash floods; many died in their automobiles. Motorists were trapped after swollen streams suddenly leaped their banks and filled roads with roiling water. Such victims little suspect that so common a phenomenon as rain can be lethal until their vehicles are overwhelmed by the awesome power of raging water. Often the roads disappear

C. J. Walker/*PALM BEACH POST*; Mike Clemmer (opposite, both); John Lopino/*PALM BEACH POST* (following pages)

Houses ground to mulch strew Homestead Air Force Base after Andrew. The pattern of destruction—utter ruin alongside areas of little damage—suggests the action of mini-swirls, meteorologist T. T. Fujita's term for vortices embedded in Andrew's eyewall.

as the pavement itself is washed away.

Alberto's flooding triggered a massive emergency response familiar to all localities that have endured disaster. From across the nation, workers—both paid and volunteer—rushed in to help those to whom Alberto had brought pain or loss. In charge was the Federal Emergency Management Agency (FEMA), the nation's field commander when nature goes on the warpath. Many of the volunteers for the American Red Cross, Salvation Army, and other groups were veterans of disasters past, such as the flooding in the Midwest that occurred in 1993.

"In times like this," observed FEMA staff member Stef Donev, "you see places at their worst and people at their best."

Winds, not flooding rains, were the weapon of choice when Hurricane Andrew flayed South Florida and Louisiana in 1992. It won—hands down—the title of most destructive storm in U.S. history.

Like Alberto, Andrew was born from an atmospheric disturbance over West Africa. Disturbances like the one that spawned Andrew, known as tropical waves, march with regularity westward into the eastern tropical Atlantic Ocean. Forecasters typically track 60 to 70 such waves during the June-through-November hurricane season.

As they continue westward, a few of the tropical waves develop concentrated clusters of thunderstorms in association with counterclockwise airflow near the ocean surface. A small number each year become hurricanes of note. In mid-August 1992 one of these tropical waves was the origin of Andrew. Hungrily it fed on the two staples of tropical storms—the heat of the sea surface and the limitless water available for evaporation.

Gaining strength as it moved westward, the storm developed a central circulation, then a chimneylike eye.

With the Frenzy Past, Shock Sets In

For two days after Andrew struck and moved on, a numbness paralyzed many of the worst hit residents of South Florida. This victim silently grieves atop the rubble of his mobile home in Homestead. Mobile and modular homes proved extremely vulnerable among the 80,000 residences made uninhabitable and the 55,000 damaged but habitable. A hundred and sixty thousand people became newly homeless. Field kitchens, tent cities, and emergency trailer parks sprang up. The last occupants left their temporary quarters two and a half years after the dreadful storm struck.

Ben Van Hook/BLACK STAR

These developments were watched with growing concern by meteorologists at the National Hurricane Center, the nation's nerve center for hurricane forecasting (located in Coral Gables, Florida, at the time but now housed in a special hurricane resistant facility on the campus of Florida International University in Miami). Like gnats attacking a charging elephant, reconnaissance aircraft of the U.S. Air Force Reserve were flying sorties into the storm to report its vital signs: pressures, winds, temperature.

On Sunday, August 23, Andrew fell on the Bahamas, pounding the islands with winds of 150 miles an hour and bringing a 15-to-20-foot-high storm surge—the great dome of water that often roars inland with a hurricane and can tower 30 feet from base to brim. Four persons lost their lives to the forces of Andrew. Earlier, at seven o'clock on Sunday morning, the "war room" at the National Hurricane Center issued an official hurricane warning for South Florida. It set in motion a frenzy of preparedness and evacuation plans, developed by the NWS and state and local officials. These safety measures significantly slashed the loss of life in the third most powerful hurricane to strike the U.S. mainland in this century.

Bringing a storm tide 17 feet high, Andrew made landfall just south of Miami at five o'clock Monday morning. It smashed through Cutler Ridge, Perrine, Homestead, and Florida City in a central swath of destruction 25 miles wide. North of this belt, forecasters at the hurricane center heard a crash. Winds that gusted to 164 miles an hour knocked out the rooftop radar antenna and also the wind-measuring instrument.

Still at full strength, Andrew churned inland, leaving 160,000 homeless in its wake. Weakening slightly, it swept across the Everglades and leveled 70,000 acres of mangroves. In two more days it had crossed the Gulf of Mexico and was thrashing out its death throes in Louisiana's bayou country, with massive destruction and a toll of 17 more lives. In Florida and in Louisiana the threat of Andrew drove a wave of more than two million people from their homes.

Trial After Tantrum

The wrath of Andrew reshaped not only the landscape but the lives of those it touched, often for long months afterward. Mosquitoes bred in standing water, rats rattled the rubble as they fed, and fear of disease spread. A Monroe County DC-3 releases a mosquito spray over a hard-hit trailer park in Homestead (opposite). Life devolved into waiting in line (below) for water, for ice, for food, for other supplies, and for the few working telephones. Assistance arrived in a swelling tide: first the Southern Baptists with their mobile kitchens, the American Red Cross, federal troops, and a host of other aid groups. Then began the long cleanup process.

Like many such disasters, Andrew left behind costly but valuable benefits: new insights into the behavior of hurricanes, new ways to reduce their damage.

Andrew was an extraordinarily powerful storm, surpassed in this century only by a Florida Keys hurricane in 1935 and by Camille in 1969. Analysis of aerial surveys of debris patterns left by Andrew in South Florida show narrow swaths, only 50 to 300 feet wide, of exceptionally heavy damage. A hundred feet from these streaks are areas of almost no damage. What fine meteorological instrument incised this precise pattern of destruction?

Possibly it was mini-swirls, perhaps only 50 feet across and embedded in the hurricane's eyewall, theorizes T T Fujita, professor emeritus of the University of Chicago and a leading authority on cyclonic severe storms. He believes these mini-swirls gained speed from the updraft of eyewall winds and the forward speed of the storm itself, until they spun as fast

Allan Tannenbaum/SYGMA; Mark Mirko/ *PALM BEACH POST (opposite)*

Chris Johns; Tom Sobolik/BLACK STAR (upper)

Broken but Unbowed

Hurricanes played similar themes of havoc, oceans apart. The Atlantic House Restaurant stood as a landmark of the good life in Folly Beach, South Carolina, before Hugo pummeled portions of the Southeast in 1989. Here it lies reduced to rubble (above)—but the day afterward, the owner vowed to rebuild. In 1992 the onslaught of Iniki battered a tavern on Kauai (opposite) but failed to close it.

as 200 miles an hour. "At that velocity everything—even Dumpsters and concrete—would be flying," he observed. "Such winds could destroy a house within a few seconds."

Can the appalling property loss to hurricanes be curbed? I found partial answers—both positive and negative—during a visit to the hurricane center in August 1994, as Hurricane Chris was sputtering out over the North Atlantic.

I called on Robert C. Sheets, then the center's director and a hands-on forecaster who has flown more than 200 scientific sorties into the eyes of hurricanes. "There's a limit to what forecasts and warnings can do," said Sheets, referring to the damage inflicted by Andrew and other hurricanes. "Most communities can't be totally evacuated, no matter how many warnings we issue. We need better construction practices, stiffer building codes, more shelters."

Acting on his concern, in the aftermath of Andrew, Sheets sat on a Florida board, convened to review construction requirements. "South Florida already has the nation's strongest codes," he observed. "Houses must be able to withstand winds of up to 120 miles an hour by use of concrete tie beams and columns. If Andrew had hit anywhere other than in South Florida, damage would have been total." He showed photographs of houses whose doors and windows had failed and that had been gutted by the winds but still stood firm.

"Among houses built to code, the main losses came from impacts of flying debris and from the failure of doors and windows; the air rushing in, coupled with the suction of outside winds on the roof and corners of buildings, caused them almost to explode. In the long run, we'll have to work through the insurance industry to bring building codes up to safe levels."

What about coastal areas lying in the

Target for Tragedy

Bangladesh stands squarely in the path of disastrous cyclones: Seven of the nine deadliest storms of the century have swept its populous, low-lying coast, claiming more than half a million lives. A 1991 cyclone, packing winds of up to 145 miles an hour, killed 139,000, mostly by the 20-foot storm surge that engulfed offshore islands. Houses, crops, and even the land itself vanished when tidal waves washed over Hatia Island (right) and *chars* built by silt from rivers that drain into the Bay of Bengal. Famished survivors on Kutubdia Island grasp for relief supplies flown in by helicopter (opposite, lower). The cyclone killed 45,000 on the island and left many thousands homeless. A young man shields his nostrils from the stench of decomposing bodies (below). Nearly half a million dead farm animals added to the health menace. To reduce casualties caused by the massive surges, foreign governments and private relief agencies are pushing to build concrete storm shelters on the vulnerable islands. In 1994 a warning system helped save many lives when a cyclone hit.

Dorio Mitidieri/SELECT PHOTOS (above and opposite); Rahman/SIPA PRESS (below)

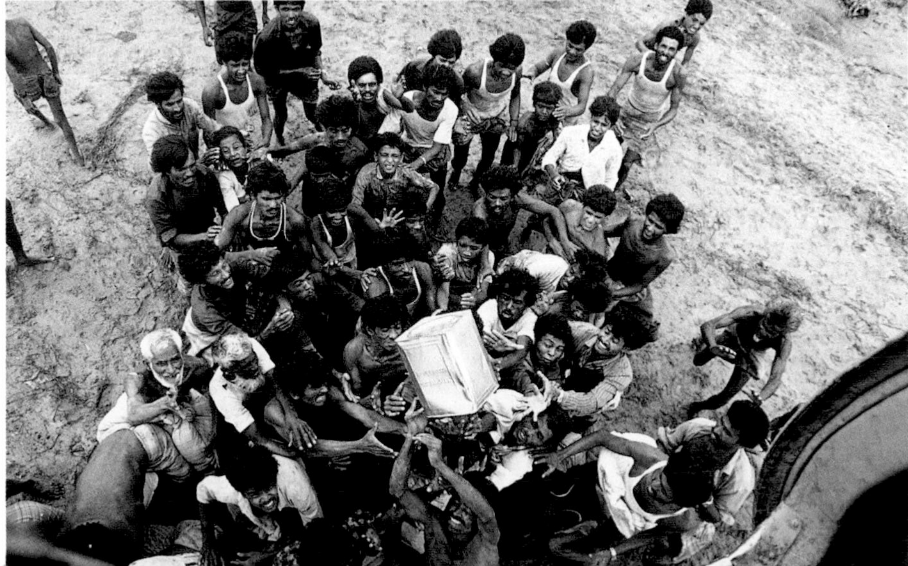

Joe Towers/WEATHERSTOCK

path of a hurricane's surge? In most hurricanes—Andrew was an exception—the surge accounts for the main loss of life.

"There's no stopping a surge; the defense is evacuation," explained Sheets. "Most coastal highways can't handle the traffic of those who must flee the surge zone in addition to those who choose to escape the winds. We need community shelters for people who want to flee the winds. This would free the highways for people escaping the surge zone. We can attempt to give earlier warnings, to provide longer lead times for safe evacuation. But the more extended the forecast, the more uncertain the accuracy."

Are those forecasts improving? I sought the answer from Hugh E. Willoughby of NOAA's Atlantic Oceanographic and Meteorological Laboratory in Miami. Like Sheets, Willoughby has spent much of his professional life flying into hurricanes.

"We improve the accuracy of forecasts nearly one percent a year," he said. "When a hurricane warning is raised, typically it applies to a 300-mile stretch of coastline. A hurricane's swath of damage will usually span 100 miles. The 100-mile cushion on each side gives room for the average error. But this error is expensive. It costs, on average, about $300,000 a mile for residents and businesses to evacuate—90 million dollars altogether. And if people evacuate and then find it was unnecessary, they are less likely to do it next time, when lives may depend on it."

Andrew formed in the Atlantic hurricane basin, which consists of the North

Tower of Trouble

The majestic sculpture of a massive thunderstorm spreads its anvil top 45,000 feet above the Great Plains. Such organized storms are known as supercells, severe thunderstorm cells that can be 5 to 10 miles in diameter. Unlike common thunderstorms that last tens of minutes, supercells can persist for several hours, travel hundreds of miles, spawn large tornadoes, heavy rains, and hailstorms. The tempests form around a rotating core and central updraft known as a mesocyclone, the source of their distinctive organization, power, and life span.

Atlantic, the Caribbean Sea, and the Gulf of Mexico. Other hurricanes that may hit the contiguous United States take shape off southwest Mexico in the eastern Pacific. Two other spawning grounds in the Northern Hemisphere breed these dreaded tempests. The most prolific is the western Pacific, including the South China Sea, where the storms carry the name "typhoon," from the Chinese *tai feng*—great wind. In the record year 1993, 32 great winds battered the Philippines. Typhoons may also lash China, Taiwan, Japan, and South Korea.

The fourth spawning ground is in the Indian Ocean. These storms, called cyclones, are fewer in number and can be extremely destructive.

Born in the Bay of Bengal off India, cyclones often move northward toward low-lying Bangladesh, driving with them the characteristic surge. The most recent major disaster stunned the world in 1991. Hurling a 20-foot-high dome of water, a surge utterly erased densely populated villages and covered the islands they sat on; 139,000 Bangladeshis died. A 1970 cyclone claimed 300,000 lives—one of the greatest natural disasters in history.

The 1991 catastrophe led to the building of more shelters and the setting up of a cyclone warning system that is credited with limiting the toll of a 1994 storm to a few hundred people.

It is Palm Sunday, March 27, 1994, and parishioners are gathering for the eleven o'clock service at the Goshen United Methodist Church in Cherokee County, Alabama. Outside, the clouds hang heavy, threatening; inside, the comfort of worship beckons. And the rural structure slowly fills.

At two in the morning of the day before, meteorologists at the National Severe Storms Forecast Center (NSSFC) in Kansas City, Missouri, had identified conditions favoring the formation of

Light Fantastic

"As if a bomb had blown up beside me—the bolt's energy knocked me down," said photographer Warren Faidley of lightning that struck an Arizona tank farm less than 400 feet from him (below). He was not complaining; lightning kills nearly a hundred persons yearly in the United States. His hard-won reward: a rare close-up view of lightning at the instant of striking. To the trained eye, it reveals the upward-moving streamers that initiate a cloud-to-ground strike. In a sunset panorama south of Tucson (right), lightning dances between clouds of a thunderstorm. Most lightning occurs within the clouds. A lightning channel has temperatures of 50,000°F, four times hotter than the surface of the sun; heating causes rapid expansion of the channel, in turn creating the shock wave heard as thunder.

Warren Faidley/WEATHERSTOCK (both)

Gene Moore/PHOTOTAKE; John Raoux/*ORLANDO SENTINEL* (opposite upper); Joe Burbank/*ORLANDO SENTINEL* (opposite lower)

supercells—large thunderstorms that breed tornadoes—over a broad area in the southeastern U.S., including Cherokee County. They launched periodic advisories to NOAA weather forecast stations in Alabama and Georgia. During the next 30 hours the stations in those states would transmit advisories over NWR, the NOAA Weather Radio that flashes continuous weather updates to those who own the inexpensive NWR receivers.

At ten o'clock that Sunday morning the NSSFC issued its Most Severe Weather Outlook, an advisory reserved for the most dangerous situations. It, too, was relayed over NWR.

By eleven o'clock 125 worshipers fill Goshen's high-roofed sanctuary. The service begins.

Five minutes earlier and two counties away, a killer tornado had touched down. It roars like a freight train toward Goshen, killing a boater in St. Clair County and a motorist in Calhoun County.

At 11:39 the tornado seizes the Goshen United Methodist Church. Haymaker winds slam into the sanctuary roof and lift it, and the walls fall down on the terrified congregants. Beneath the rubble, 20 people die; another 90 lie injured.

A total of 18 tornadoes rampaged through Alabama and Georgia on that Palm Sunday, killing 42 people and destroying property valued at 107 million dollars. Second only to the Goshen Church tragedy in magnitude of loss was the devastation visited upon residents of mobile homes throughout the area. The sustained power of the supercells that spawned these twisters astonished meteorologists. A canceled bank check lifted near Goshen was found 130 miles downwind in northeast Georgia.

Worse tornadic sprees have lashed middle and eastern America in the past. Lurid photographs of their devastation

Hail's Brief but Costly Barrages

Skies blacken, lightning flashes, rain pours—and suddenly fusillades of ice pellets crash down on cars, buildings, and humans. Vehicles yield the road to hail near Quail, Texas, in 1977 (opposite). Hailstones blanket Orlando, Florida, in 1992, clogging streets (above) and riddling car windows (below). Dubbed the "white plague," hail in an average year inflicts damages to crops and property amounting to nearly a billion dollars. One hailstorm in Texas in 1995 resulted in a billion-dollar loss. The week before that, a smaller hailstorm did 250 million dollars' damage.

E. DeWayne Mitchell (all); NWS/WEATHERSTOCK (following pages)

decorate the main corridor of the NSSFC. "The pictures sort of rev up the incoming shift for the day's work," said deputy director James Henderson.

The centerpiece of this meteorological gallery was a montage labeled "10 Famous Outbreaks." First honors went to the "Super Outbreak" of 1974, in which a barrage of 148 tornadoes peppered a dozen states centered around Kentucky and Tennessee: 315 dead, 600 million dollars' damage. Close behind came the infamous Tri-State outbreak of 1925, with its appalling toll of 689 killed in Missouri, Illinois, and Indiana.

Henderson pointed to charts showing tornado frequency and fatalities for successive decades. "You can see that the number of tornadoes is soaring—from 1,685 tallied in the 1930s to more than 8,000 in the 1980s. This is probably because of better reporting. You see also that the number of deaths is plunging—1,947 in the 1930s compared to 585 in the '80s. This drop probably results from better education, preparedness, warnings, and our tornado alerts."

Tornado and severe thunderstorm alerts are the primary business of the center's Severe Local Storms Unit. Banks of humming electronics flickered with satellite images, data from the hundreds of weather balloons released twice daily at airports, a real-time recording of lightning flashes across the nation, inputs from mathematical forecast models, and information gathered from the National Weather Service's vast array of radars at airports and other facilities.

How successful is the unit in warning about killer tornadoes? With lead forecaster Jack Hales, who drew duty that fateful Palm Sunday, I reviewed the Tornado Fatalities Record for the first six months of 1994. "We show 16 killer tornadoes, with a total of 55 fatalities," Hales noted. The NSSFC issued official tornado watches on all but two.

Forecasters benefit vastly from improved radar known as Doppler. Conventional radar sees precipitation; Doppler portrays the invisible winds and measures their speed and direction. This vision greatly enhances meteorologists' ability to peer inside churning supercell storms and detect rotation within the clouds and, perhaps, hidden tornadoes.

"We're seeing disturbances now in clear air—things we didn't know existed," said NOAA's Joseph Golden, veteran of 20-plus years of storm chasing. "We see miniature cold fronts that converge and cause severe storms. We've discovered

that urban areas create their own cyclonic airflows and that these can influence the location and strength of storms. Doppler enables us to predict the amounts of rain that will fall and warn of some flash floods; it shows us the onset of hail and the changes of wind speed with height, which indicates where severe disturbances will occur on the ground."

But a crucial limitation plagues all radar: Because of the curvature of the earth, it cannot see distant weather at ground level. Partly for this reason, scientists have been unable to discover how tornadoes form before working their destruction. If this mechanism could be identified, perhaps its early stages could be detected at heights radar can see, which would help in tornado forecasting.

This is a goal of a tornado-chasing program hosted by NOAA's National Severe Storms Laboratory (NSSL) in Norman, Oklahoma. Called VORTEX, the program sends vehicles bristling with meteorological probes across the prairie in hot pursuit of tornadoes, while overhead balloons and aircraft equipped with Doppler radar spy on the storms, unimpeded by earth's curvature.

Doppler radar has already given scientists insights into the deadly phenomenon of downbursts. They are the violent force that causes planes taking off or landing

Making of a Monster

A tornado born of a supercell grows to full fury near Laverne, Oklahoma, in 1991. Seen from about three miles away, the genesis takes place near sunset over a time span of nearly ten minutes. As with most tornadoes, the twister takes on the coloration of the debris its funnel vacuums up—here the red soil and vegetation of prairie rangeland. The storm wielded winds estimated at 180 miles an hour and hailstones the size of softballs. Passing south of town, it overturned a house trailer and destroyed a five-room house. Three people were hospitalized. Frequent collisions of warm, moist air and cold fronts make the central U.S. the world's capital for tornadoes. They stir the fastest surface winds known, reaching 300 miles an hour. Texas, Florida, and Oklahoma record the largest number, with Mississippi suffering the most fatalities from tornadoes—on average, ten a year.

A tornado touches down near Enid, Oklahoma, in 1966, unroofing homes, overturning boxcars, and generally creating havoc. Thousands watched, but no one was injured.

to crash suddenly on or near the runway.

Two costly disasters within three months in 1975 gave impetus to studies that revealed the identity of these vicious winds. In June Eastern Airlines Flight 66, as it approached for a landing at Kennedy International Airport in New York, crashed on the runway as a result of unknown causes, killing 113 people. In August, Continental Flight 469, taking off from Stapleton International Airport in Denver, slammed down in a wheat field near the runway, injuring 15. Both events occurred during thunderstorms seemingly safe to fly in.

Painstaking reconstruction of the thunderstorm patterns led meteorologist T. T. Fujita to identify the presence of small but powerful jetlike downbursts. Later they became known as microbursts, a name coined by Fujita. When a microburst's violently descending column of air contacts the ground, the impact forces the wind upward in a swirling vortex. Microbursts may be wet or dry, and—by definition—they extend little more than two miles over the earth's surface. Although they last only a few minutes, they are extremely dangerous.

Recognition of the microburst problem and the wider adoption of Doppler radar for detecting suspicious storm patterns will undoubtedly reduce air losses—losses like the one in July 1994 when a USAir passenger plane, landing without Doppler at Charlotte, North Carolina, crashed. The death toll was 37. During the mid-1990s, 47 storm-prone airports are scheduled to receive a new generation of Doppler designed for microburst detection.

Those who fly into weather's realm encounter other threats: turbulence, poor visibility, icing, even volcanic plumes. For warnings, pilots and navigators depend on forecasts from the National Aviation Advisory Unit in Kansas City. Around the clock, meteorologists toil over their

Reaping a Whirlwind

Like an angry beast, a tornado roils north Texas farmland, two miles away from University of Oklahoma graduate students. Bracing a portable Doppler radar unit, their goal was to learn why some thunderstorms produce twisters and others do not. For the first time, a chase team was able to set up and use a portable Doppler radar within close range of a tornado. The ultimate goal: to improve tornado prediction. Accurate prediction failed to avert disaster when a flurry of 18 tornadoes raked Alabama and Georgia on Palm Sunday 1994. One twister lifted the roof of the Goshen United Methodist Church in Piedmont, Alabama (below). Collapsing walls killed 20 of the worshipers that fateful morning and injured 90. Tornado advisories had been issued but had gone unheard.

AP/WIDE WORLD PHOTOS; Howard B. Bluestein (opposite)

electronic tools to divine the perils lurking in the wide belt of atmosphere reaching from ground level to 45,000 feet, the usual ceiling for commercial aircraft.

"The greatest hazard," explained duty forecaster Paul Smith, "occurs when visibility drops to less than three miles, or the ceiling to less than a thousand feet. Then pilots must fly on instruments."

Some areas are particularly hazardous. The Rockies and other mountains cause turbulence. Dry air coming off the Rockies creates conditions for microbursts. The Cascades are bad for icing—a veritable icing machine—as are the Great Lakes, where heavy snow showers are often the culprit. In addition, freezing drizzle can coat a plane so fast it sinks to the ground before the pilot can fly out of it.

"We try to foresee all of these," said Smith. "The need for safety in the skies has been a major stimulus to weather forecasting generally."

The estimated 100,000 thunderstorms that lash the United States annually result in 20 million lightning strikes to the ground. A single storm often produces the energy of a small nuclear power plant. Of those 100,000 thunderstorms, the NWS classifies 10 percent, or 10,000, as "severe"—producers of hail at least three-quarters of an inch in diameter, winds of at least 58 miles an hour, or tornadoes.

Thunderstorms and their accompanying

Where Twisters Trod

Ruins strew two Texas towns visited by tornadoes. Trees as well as houses splintered when a 1994 storm struck Lancaster, a Dallas suburb (above). Saragosa, a Hispanic village, lies virtually erased in 1987, when 80 percent of its buildings were shattered (right). Two-thirds of the 30 fatalities occurred at a pre-school graduation ceremony. Winds flung a fork 100 yards into a tree (left).

David Sans/SIPA PRESS; Warren Faidley/WEATHERSTOCK (opposite and below)

When Planes Plunge

During thunderstorms, aircraft taking off and landing risk losing lift from wind shear, often in the form of microbursts. A microburst is a vertical column of air more than a mile wide and up to 25,000 feet high. In this storm near Rutland, Illinois (right), plunging cold air splashes outward, stirring the rotating dust column at far right. Low-level wind detectors installed in the mid-1980s help warn of wind shear and reduce the problem. Sophisticated Doppler radars, for example NEXRAD and TDWR, appeared in the early '90s, but installation often lagged, as at the Charlotte, North Carolina, airport. In July 1994 a microburst may have contributed to the downing of USAir flight 1016, landing at Charlotte (below). The jet's tail section ploughed into a house. Of the 57 travelers, 37 died in the crash.

Ronald D. Brewer; Craig Bell/*GASTON GAZETTE* (opposite)

lightning were pounding assorted parts of the nation (a normal day's weather) when I visited the NSSL in Norman. The violent panorama unfolded before me on a map displayed on a computer monitor as I sat with Ron Holle, an NSSL lightning expert. Flash!—a strike from a storm climbing

Steve Northup/BLACK STAR (both)

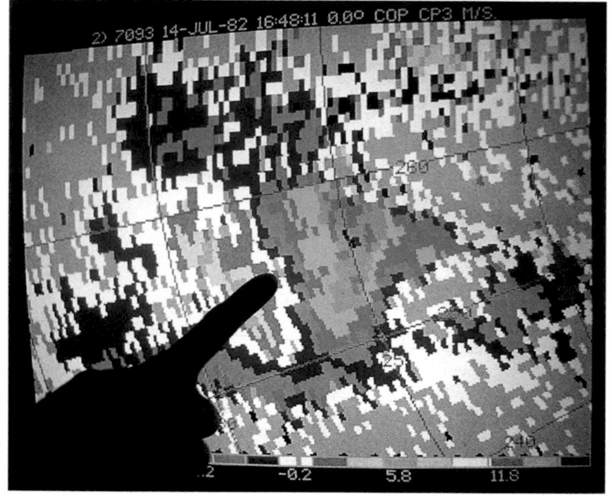

Arizona's Mogollon Rim. Flash! Flash!—turbulence in the Rockies. Frenzies of flashes!—a mighty storm system pummeling Missouri, Iowa, and Illinois.

"Ground sensors placed throughout the continental U.S. record the strikes and transmit signals to a satellite, which returns them via a receiver in California to Tucson, Arizona," explained Holle. "The system helps electrical utilities save tens of millions a year by telling repair crews where to look for causes of power outages—and where to insulate against lightning. It tells foresters where to look for lightning-caused fires."

Each year lightning-caused fires and other damage cost Americans as much as a billion dollars. Holle believes this figure will climb.

So could the death toll.

In June 1994 a powerful thunderstorm moved quickly over Lake Moomaw in western Virginia. A family of five, fishing on the lake, hurriedly made for shelter on a small island. Next day a fisherman saw their boat still there and investigated. A bolt of lightning had struck a tall pine and split it to the ground. The bodies lay in a 15-foot radius from the tree, all burned, all killed instantly.

Group casualties are the exception with lightning; usually it picks off its victims one by one.

An increasing frequency of lightning in thunderstorms can mean danger in many forms. Farmers know it can presage hail.

It is late August in North Dakota, in the rich farmland just west of the Missouri River. All afternoon Leland Graner and his brother Eugene, both in their 60s and as close to one another as their contiguous farms, watched clouds build to the west. "They looked pretty benign, at first," recalls Eugene. But the brothers worried. For two days the NWS had warned of the possibility of severe thunderstorms, and maybe hail. Worse, the corn was almost ready for harvesting.

"About five o'clock the sky changed," recalls Eugene. "A big, black cloud began building, fast. Within 15 minutes it hit. The wind drove rain and hail horizontally—hail the size of golf balls."

In 25 minutes it had passed. The men inspected the damage. Windows were broken in both Graner homes; vegetable gardens, shredded; corn stalks, stripped of leaves or smashed to the ground.

I arrived a week later to find Leland Graner salvaging the remains of his crop. His tractor pulled a chopper to shred ears and stalks into silage. Eugene and I followed in a second tractor, pulling a dump box to relay Leland's silage to the silo. "The disaster hit us hard," Eugene told me. "It's rough on a man. He works the soil, fertilizes, buys the seed, plants it, irrigates two or three times—a lot of work, at all hours of the day and night. He brings it along until it's ready to harvest—and a few minutes of hail ruin it."

Losses like the Graners', repeated thousands of times across the nation, add up to annual losses of nearly a billion dollars from hail damage to crops and property. The costliest hail disaster hit Dallas–Fort Worth on May 5, 1995, when softball-size missiles shattered windows of cars and buildings. Insured damages could amount to a billion dollars.

Can these enormous losses be reduced? A NOAA/state cooperative atmospheric modification project in western North Dakota offers some hope of being able to change weather.

"We call it the North Dakota Cloud Modification Project," explained director Bruce Boe, in his office near the state capitol in Bismarck. "When our people detect the buildup of thunderstorms, we throw as many as eight aircraft skyward."

If the storm is a potential hail-maker, a cloud-seeding plane flies into the forward part, where cloud turrets are bringing up moisture that fuels the storm's growth. Seeding the turrets with microscopic salty silver iodide particles or dry ice accelerates the formation of rain. The purpose is not only to reduce the formation of hail but also to increase rainfall.

In the five counties that support the project, operating costs of about half a million dollars a year produce about 19 million dollars in benefits from the reduced wheat loss and the increase in yield from more rain.

Satellite meteorologist Hank Brandli stares in disbelief at his home weather-satellite receiving system, tuned to polar orbiting weather satellites 450 miles up.

On the computer monitor, an enormous storm, part of which is battering his Florida home, engulfs the entire East Coast with snow, ice, fierce winds, tornadoes, torrential rain, and a murderous storm surge.

His awe is justified. The date is March 13, 1993, and the monster he sees will win the title of Superstorm of March

To See the Wind

Doppler radar provides electronic eyes for detecting wind shear. Electrical impulses reflected by wind-borne rain, snow, hail—even insects—up to 30 miles away enable Doppler radars to measure wind speeds and direction. Research weather radar, housed in a dome near Boulder, Colorado, (above) played a key role in developing the system concepts for NEXRAD. A danger zone for aircraft shows up in a Doppler image of a storm (opposite). Divergent outflows from a severe downdraft of a microburst appear as the closely spaced blue-green area of inbound wind and the yellow-brown area of outbound wind.

Storm of Historical Magnitude

The National Weather Service predicted a "storm of historical magnitude" 48 hours in advance. It was, indeed, a superstorm that bludgeoned eastern North America from Cuba to Canada in March 1993. The monster unleashed storm surges, tornadoes, smothering snows, torrential rains, and paralyzing cold that broke or tied 140 records. Assaulted by snow, sleet, and rain, New Yorkers (above) also contended with power outages. Wind-churned waves washed 18 homes in Southampton on Long Island (below) out to sea. In the third day of the rampage, a satellite image (opposite) shows the northeaster still punishing Maine and eastern Canada.

AP/WIDE WORLD PHOTOS; Rick Maiman/SYGMA (upper); Hank Brandli (opposite)

1993—sometimes erroneously referred to in media reports as the "storm of the century." Dwarfing the mightiest hurricane, for three days it rages from the Gulf of Mexico to New England and Labrador. Ultimately it will be responsible for some 200 deaths and the destruction of property worth more than two billion dollars.

Florida, still bleeding from Hurricane Andrew, feels the storm's full fury. Northward along the entire Atlantic coast and inland beyond the Appalachian Mountains, the enormous storm spreads destruction: beachfront property smashed, crops and forests damaged, roofs collapsed by snowfalls of three feet, electric power lines downed by high winds. Vast regions paralyzed.

When its fury finally is spent, the superstorm has affected the lives of a hundred million Americans. It was a classic northeaster: frigid polar air sweeps down the Mississippi Valley, enters the Gulf of Mexico, picks up huge amounts of moisture, then turns northeastward along the East Coast, dumping snow in winter or rain in summer.

Paradoxically, where winter storms dump snowfall, a major cause of death comes after the storm's fury has passed and the cleanup has begun. For example, in Pennsylvania, where an estimated 52 people died from the Superstorm of March 1993, only 4 succumbed during the event—from exposure. The remaining 48 victims died of heart attacks while shoveling snow.

Severe storms can strike swiftly and take many forms, but most observe a welcome ethic: They depart almost as quickly as they come.

Not all bad weather is so obliging. At times the world's familiar climate patterns are reshuffled, as during an El Niño. Then regions and even continents struggle to adjust to adverse weather that seemingly has no end.

A Young Life Falls to an Ancient Enemy

Salvaged from the ruins of the family's mobile home, a portrait of six-year-old Angie Register poignantly personifies the human tragedy of weather on the rampage. She died, and four other family members were injured, in one of five strong tornadoes that ravaged Florida during the March 1993 superstorm. In all, the massive event caused more than 200 deaths, inflicted damages amounting to more than two billion dollars, and affected the lives of at least a hundred million people, ranking it among the worst storms of the century.

Timothy O. Davis/*GAINESVILLE SUN*/SILVER IMAGE

Climatic Events

by Leslie Allen

Baby-sitting grandparents flee a 1994 flood in Conroe, Texas, with their grandchild and the family dog.

The old town is dead, though not yet demolished. Here, along its main street, stands Farmers' State Bank, a tidy brick building; a few doors down, Valmeyer Lumber Supply. White frame houses march off along side streets. A weathered Monte Carlo slowly rolls in the direction of the levee. Other than its driver, there's not a soul in sight.

At first glance, the town could be just another slumbering has-been in the land of new malls and subdivisions. But a closer look reveals a more violent fate, as backyard piles of mangled furniture, collapsed roofs, and scattered sandbags come into focus. In a living room, a ten-foot length of linoleum flooring rides a sofa, and objects and furniture appear to have been distributed around the room by a whirling centrifuge.

When the Mississippi River rolled through Valmeyer, Illinois, on August 2, 1993, it heaved the Corner Pub's bar atop several barstools and muscled a six-foot table from one room into another. The pub's owners, Bernice and Bill Meadors, discovered this several days later when Bill Meadors returned by boat. "I just stepped out on the roof, kicked in a window, and went inside," he explains. They'd owned the tavern less than two years. Now it was beyond repair.

"You don't think anything that bad can happen," says Bernice Meadors. "It happens somewhere else."

And it had—at least since 1947, when a foot of Mississippi River water had lapped at Valmeyer's baseboards. That was before the Army Corps of Engineers built a formidable double levee system on the riverbanks north of town. During the summer of 1993, as reports of catastrophic flooding and record-breaking river crests upstream reached Valmeyer, old-timers who remembered the '47 flood simply moved their belongings upstairs.

Steve McCurry/MAGNUM; Paul Chesley (opposite); David Leeson/ *DALLAS MORNING NEWS* (preceding pages)

They never bargained for floodwaters 16 feet deep. They never bet on one of the worst weather disasters in U.S. history draining the lifeblood out of their town of 900 residents.

"It was a roar like a train coming in the distance." Waiting for the inevitable in a police cruiser, Mayor Dennis Knobloch heard the oncoming river pouring through the breached levee in pre-dawn darkness. But the violence with which the Mississippi finally reconquered its floodplains belied the slow, subtle gathering of long-term weather conditions leading to the disaster.

Though floods rank as the most prevalent natural disasters in the United States, most have a clear genesis in a particular short-term weather event. They include notorious American disasters like the Johnstown, Pennsylvania, flood of 1889. More than 2,000 people died after an intense rainstorm flash flooded the Conemaugh River and rising water pressure melted the earthen dam 15 miles

Too Much, Too Little

A run of extreme weather events in the 1980s and into the '90s continues to provide baffling new data about climatic systems. Some disasters have resulted from too much of a good thing—India's rain-bearing monsoon winds, for instance, which relieve the sweltering subcontinent each summer. Opposite, Varanasi's streets become waterways as the rain-swollen Ganges overflows its banks during the 1983 monsoon. Elsewhere in the world, the extreme has been too little moisture. Although drought is no stranger to arid Australia, a four-year Big Dry that began in Queensland has destroyed all records. Above, a sign by a dried-up lake recalls predrought days.

After Months of Rain, Levees Give Way

Records, along with levees, crumble as the Great Flood of 1993 visits destruction upon the American heartland. Among the 1,100 levee breaks along the Mississippi and its tributaries, one of the most catastrophic occurred near Valmeyer, Illinois (opposite). The town, four miles away, was submerged under 16 feet of water. Downriver, historic Ste. Genevieve, Missouri, fared better, though floodwaters damaged hundreds of homes. There, more than 750,000 sandbags, including the thin white line ringing one home (below), helped to blunt the water's force.

Andrea Booher/FEMA; Jeff Christensen/REUTERS/BETTMANN (opposite)

Paul Childress; Jodi Cobb, NATIONAL GEOGRAPHIC PHOTOGRAPHER (opposite)

upstream from town. Worldwide, flood disasters include staggering catastrophes like the drowning of 139,000 or more Bangladeshis after a cyclone in 1991.

No one can say just when the Great Flood of 1993 really began. Was it in July 1992, when farmers and hydrologists noticed particularly soggy soil in the upper Mississippi Valley? Or November, as heavier-than-normal rains began a relentless drumbeat across the region? Or April through June 1993, the wettest spring the area has seen since record-keeping began in 1895? The wet spring gave way to a wetter summer, and saturated fields yielded huge crop losses. Rain continually fell on waterlogged soil, and turgid rivers began to fill up. The first of some 1,100 levee failures occurred on June 20, Father's Day; by the Fourth of July, some 100 rivers had overtopped their banks. In the end an estimated 50 people lost their lives, and the flood caused some 10 billion dollars' damage throughout the Midwest.

Debating the finer points of Great

Death Revisited

National Guardsmen recover a decades-old steel vault from a watery grave downriver from Hardin, Missouri. Magnifying the misery of a flooded town's residents, the Missouri River turned ghoulish, scooping a 65- to 70-foot-deep crater under Hardin's 180-year-old cemetery and redepositing caskets in tree limbs, railroad tracks, and bean fields as far as three towns away. All the more shocking: Not even a river town, Hardin lies five miles north of the Missouri. Where the Des Moines and Raccoon Rivers rushed on a collision course, even round-the-clock sandbagging (opposite) could not outstrip cresting waters and thwart massive flooding in Des Moines, Iowa. Adding irony to insult, the rampaging Raccoon flooded the water treatment plant, leaving the waterlogged city of Des Moines without drinking water for 19 days.

Flood weather patterns will keep meteorologists tapping out e-mail messages for years. But one rule of thumb that even nonscientists can readily grasp is that, like many major disasters, the Great Flood resulted from not one, but several, causes.

An extremely strong Bermuda High—a high pressure system named for its usual location—moved over the Southeast and refused to budge. While the South and mid-Atlantic baked under cloudless skies, clockwise winds around the high pressure area helped warm, moisture-laden air from the Gulf of Mexico move into the upper Midwest. There, it collided with cool, dry Canadian air shipped in by the jet stream, steering far south of its usual trajectory. The excessive rains that the collision spawned couldn't be blown eastward by prevailing winds because the Bermuda High blocked the way.

Even without such an unusual chain of events, what we consider "average" weather for a given season depends upon atmospheric variables like heat, moisture, winds, and air pressure achieving an exquisite, and all-too-fragile, balance. Imbalance, in fact, drives the global weather machine because the sun's energy heats the earth and its atmosphere unevenly. Still, the Great Flood was one of many examples of extreme weather during the 1980s and '90s that made many people—including climate researchers—wonder whether what the lay person might call "normal" weather had become a thing of the past.

In the United States alone, the period was characterized by particularly sharp climatic swings. For example, in 1983 "wet-spell conditions" officially applied to more land area than during any other year between 1895 and 1994. At the other extreme, 1986 proved to be the driest year in over a century for parts of the Southeast. The Midwest's 1988 drought took a much higher human toll than its Great Flood of 1993, and California's six-year drought proved to be

the longest in the state's history.

Other extremes: The winter of 1994 struck the nation east of the Rockies with 16 major snow and ice storms during three months of numbing cold. As far south as Mississippi, ice brought down thousands of miles of power lines. In Washington, D.C., the federal government followed power company guidelines for saving energy and closed down for a day. Throughout the paralyzed freeze zone, 130 deaths, mostly due to car accidents or heart attacks while shoveling snow, occurred. Less than a year later, it was California's turn—for torrential rains, mudslides, flooded rivers, and record-high temperatures.

When it comes to extreme weather in North America, "the jet stream is often the smoking gun," says climatologist Daniel J. Leathers of the University of Delaware. This permanent high-altitude, high-speed, hundreds-of-miles-wide wind flows around the globe. "We understand quite well," Leathers says, "how the jet stream affects our weather. Basically, if you're north of the jet stream, temperatures are below normal; if you're south of it, they're warmer than normal. The jet stream itself is located at the frontal boundary; that's where storms develop."

Snaking down from the north, the jet stream and its changing patterns have been linked to much of the recent spate of unusually wayward weather. "Before the late 1950s," Leathers explains, "the jet stream tended to move primarily west to east across North America. But since then, and until the present, there have been a lot more north–south variations." With each new squiggle, areas of cold, warmth, rain, and snow are rearranged on the map.

What has caused this dramatic change? "The jet stream may indeed be like a smoking gun for extreme weather," Leathers repeats, "but that doesn't tell you who pulled the trigger." Nevertheless, there are a number of suspects. Volcanic eruptions, for instance, can tweak the jet stream, as Mount Pinatubo's eruption did in 1992. So can something as seemingly remote as the snowpack in Eurasia. Variations in climate depend on such teleconnections, distant linked causes that trigger events perhaps half a world away.

By far the most important teleconnection to the jet stream—and much of the world's weather—turns out to be the shift-

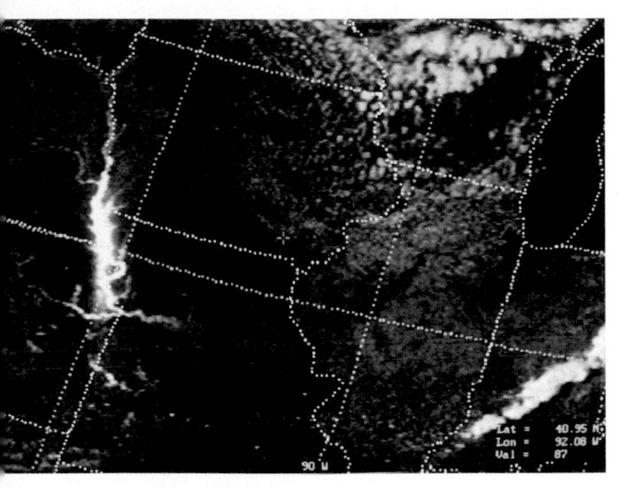

Flood of the Century

The church on Kaskaskia Island, Illinois, still rises above floodwaters (below). On July 22, the steeple bell tolled a last warning as the levee burst. Flooding created the island in 1881 when the Mississippi shifted course. Sunlight glinting off water (left) reveals the extent of flooding in the upper Midwest, as seen by a U.S. polar-orbiting weather satellite 450 miles high.

Rick Rickman; Hank Brandli (above)

Andrea Booher/FEMA (above and opposite); Rick Rickman/MATRIX; upper

Homes, Livestock, and Livelihoods Imperiled

Noah's ark yields to a Good Samaritan's outboard for a modern-day deluge (left). With livestock the mainstay of many farms, saving cattle, hogs, and sheep was a high priority. Barn roofs, railroad trestles, and church altars, along with boats, offered refuge. Kaskaskia Island farmers crowded their hogs onto the last dry spot on their ruptured levee. Street

ing sea-surface temperatures (SST) in the tropical Pacific Ocean. Where ocean temperatures increase, so do rates of evaporation, convection, and precipitation: The heat release triggers changes in the atmosphere's circulation. "When the warmer sea-surface temperatures move around a little bit," Leathers explains, "jet stream patterns across North America change dramatically."

What often nudges SST changes is the climatic juggernaut that goes by the deceptively innocent name of El Niño—the Child. Peruvian fishermen coined the name long ago to describe a warm current that flows southward along their coast every year just after Christmas. When it does, fish disappear and prevailing winds falter. The phenomenon generally wanes after several weeks.

But every three to seven years, the ocean current becomes especially warmer and stronger, and El Niño turns into a problem child. Slabs of warm water hundreds of feet thick pile up off the coast, disrupting the normal upwelling of cold, nutrient-rich water. Fish—and the seabirds that rely on them as food—die by the millions. Fishermen are idle for months. El Niños also bring devastating floods to areas far inland in Peru and Ecuador.

For more than four centuries after conquistador Francisco Pizarro's clerk described the impacts of weather anomalies that today we attribute to El Niño,

signs (above) measure the high water mark in Grafton, Illinois—more than 20 feet above flood stage. Near the Mississippi and Illinois's confluence, the town flooded in April. Receding waters brought mixed blessings, even when homes survived: Foul-smelling muck coats a kitchen in Alexandria, Missouri (right). Outside, mud-covered streets await plows.

Jeffrey Z. Carney (both); *HOUSTON CHRONICLE* (following pages)

Town Drowned, Reborn

"People have likened what happened to our town to a death, but there won't be closure while the homes still stand," says the mayor of Valmeyer, Illinois, Dennis Knobloch, seated amid the remains of his own home. Floodplain regulations barred funding for rebuilding, but offered buyouts. "That left us with two choices," Knobloch says: "Move the town to elevated ground or go our separate ways, in which case there wouldn't be a Valmeyer anymore." Two to one, residents chose the former option. But speed was essential to hold the town's citizens together, and a cornfield 400 feet above the old town was purchased in the fall. The mayor, a former banker, knew how to work with funding agencies for necessary financing. By late 1994, the new town's first homes near completion (opposite.)

they were known to weather-watchers and dismissed as local problems. Finally, by 1966, new oceanographic data showed that the abnormally warm waters of an El Niño could span the entire eastern tropical Pacific in an east-west band along the Equator.

As long as the westward-blowing trade winds hem in the Pacific's warmest waters, they pile up in the far western Pacific. There, off the coasts of Indonesia and northern Australia, the tropical weather machine brews up a roiling cauldron of rising vapor and pounding rainstorms. For El Niños to get under way, the trade winds must weaken and allow warm water to spread eastward across the Pacific. When that happens, the convective cauldron also moves eastward, shifting low air pressure to the mid-Pacific. High pressure replaces it near Indonesia in a reversal called the Southern Oscillation. High pressure brings rainless skies, while low-pressure systems are associated with cloud formations. Together, El Niño's ocean warming and the sea-level pressure pattern, the Southern Oscillation—ENSO, for short—

can deflect normal weather systems thousands of miles off course, often for more than a year at a time.

As if to test the newly formulated hypothesis about ENSO—that is, that all ENSO events follow the same pattern of development—one of the most extreme El Niño events appeared, wreaking havoc on five continents during 1982 and 1983. A few million acres of Indonesia's rain forest burned, the monsoon season never arrived to cool southern India and Sri Lanka, and crops withered under the heat from cloudless skies that stretched from the Philippines to Botswana.

Meanwhile, six tropical cyclones—an unprecedented number—battered the Pacific paradise of French Polynesia. Water vastly warmer than normal moved northward toward the West Coast of the United States, bringing exotic tropical marine life with it as far north as Washington State. In Peru and Ecuador, floodwaters spread disease. Before it dissipated, that ENSO cost the world more than eight billion dollars and claimed upwards of 1,100 lives.

Still, past El Niño events have apparently inflicted even greater human misery. In 1878, as drought ushered famine into China, an Englishman in Shanghai reported: "The people's faces are black with hunger; they are dying by the thousands upon thousands. Women and girls and boys are openly offered for sale to any chance wayfarer...." At least nine million Chinese perished; eight million died in India's famine the same year. A decade later, ENSO led to the deaths of another million or more Indians; Ethiopia may have lost a third of its entire population.

As with any kind of recurring natural disaster, El Niño's destructive wake varies from one episode to the next. But some areas are more susceptible to its ravages, time and again. "All places where El Niño operates strongly have much more climate variability, many more severe droughts and floods, than other places," says Australian (Continued on page 158)

In a heartrending homecoming, a Kingwood, Texas, couple views the damage done to their home by the rain-swollen San Jacinto River in October 1994. Ten thousand people were evacuated.

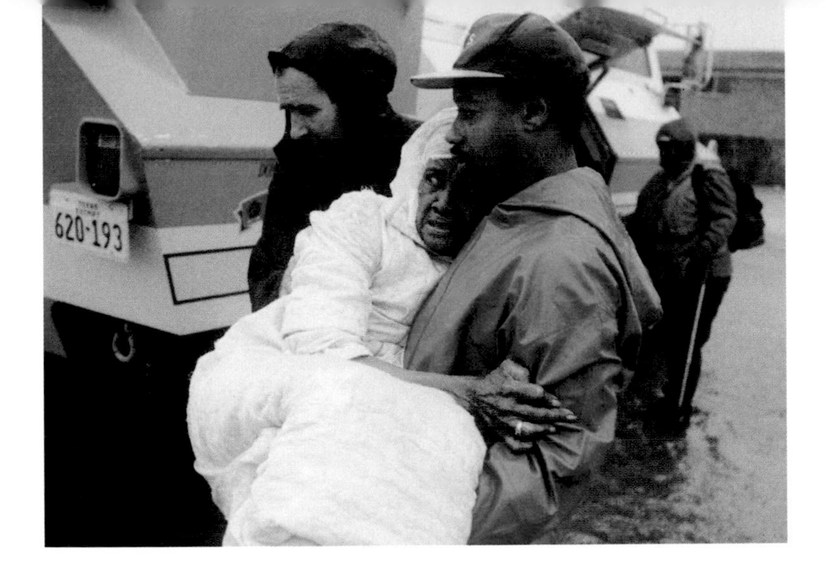

Houston's Rains Trigger a Fiery Catastrophe

Black clouds of burning fuel billow skyward as the San Jacinto River becomes a torrent of fire (opposite). Houston's natural disaster spawned a man-made one when pipelines failed and gasoline, crude oil, heating oil, and natural gas poured from them and ignited. The fire spread at more than 80 miles an hour and engulfed woods, houses, boats, and barges. Lost to the juggernaut: the home and business of William Smith, applying for loans at a makeshift Small Business Administration office (below). For 88-year-old Rosa Gregory, aid takes the form of a helpful neighbor and SWAT officers equipped with an armored amphibious tank (above).

HOUSTON CHRONICLE (all)

meteorologist Neville Nicholls, a leading ENSO researcher. "ENSO amplifies that variability." His own homeland is a prime example. "El Niño's effect on Australia is as great as it is anyplace else," Nicholls points out. "Most droughts here are associated with El Niño."

Most native plants and animals evolved rhythms of growth and fertility that follow the cycles of El Niños, according to Nicholls. But for European settlers, the consequences have proved devastating, ever since drought almost wiped out the fledgling colony in 1791.

"Across south Australia you can see the remains of old abandoned farms," Nicholls notes. "There have been many times when a wet year or two has led people to settle dry areas. Then along comes a really good El Niño, and it destroys everything." Still, nature's forgiveness, like mother love, has shone through after even the most destructive Niño has barreled through. After a year or so, the Southern Oscillation swings back to normal and rains return; usually,

HOUSTON CHRONICLE (both)

Ruptured Pipelines

Fuel from ruptured pipes burns along the San Jacinto River (below) in Houston's "spaghetti bowl," one of the nation's interstate pipeline networks. The Coast Guard estimated that 1.5 million gallons of petroleum products spilled. Booms trapped some of the oil, which spread as far as the Upper Galveston Bay. Cleanup crews beat back slicks with chemical sprays (right).

it swings so far that opposite conditions give birth to La Niña—the Girl—that spreads cold water across the Pacific and brings flooding rains to eastern Australia.

Until recently, that is. "We've had too many boys and only one girl in the last two decades," says Nicholls. "Things have been quite strange and we don't know why." Starting in 1991, three "boys," following one after another, had kept rain from eastern Australia for more than four years by January of 1995. A run-of-the-mill dry spell had given way to the Big Dry, the worst drought in recorded Australian history.

In January 1994 near Sydney on the southeastern coast, forest fires ignited drought awareness. The fires destroyed more than 200 houses and other buildings, killed four people, and burned 1.9 million acres.

To the north, across the rolling black-

soil hinterlands of Queensland, things were even grimmer. There the drought—like the great Mississippi flood half a world away—had no defined beginning. Its onset was slow and insidious, as the blazing sun sucked life-giving moisture from the soil day after day, month after month, and—finally—year after year.

For a farmer or grazier there were milestones: the day he shot a thousand sheep he couldn't feed; the day fourth-generation farmers turned out the lights and drove off. But for the most part, drought is a disaster in slow motion.

When you look, for the first time, at a land in its fourth year of drought, you see only the sum total of this slow torture. Farmland as knobbed and gullied as an old washboard gives no hint of its reputation as some of the world's most fertile, though puffs of hot breeze still loft gauzy curtains of topsoil. In midsummer the bare trees bring to mind a northern winter, but these gums have died.

Whole forests, fueled by their oily eucalyptus leaves, have burned, trapping hundreds of koalas in their upper branches. Where trees survive to spread sparse shade, clusters of gray kangaroos move

Mike Magnuson; John Van Hasselt/SYGMA (lower); AP/WIDE WORLD PHOTOS (opposite)

Storms Wreak Havoc

A blizzard roils a Minnesota town (above), and a pileup snarls interstate traffic near Milwaukee (opposite) during 1994's notorious winter. Brutal cold shut federal buildings in Washington, D.C., for a day, and snow blanketed New York (below). Sixteen major snow and ice storms struck many areas east of the Rockies, breaking snowfall and low-temperature records.

El Niño, Down Under

Soil rendered useless by drought sifts through the hands of Queensland farmer Chris Henning, unable to harvest a crop for three years in a row. Jayne Shatte comforts an orphaned lamb. Years earlier, graziers unable to sell or board their stock were forced to shoot thousands of animals. The expense of trucked-in water and feed (right) for reduced herds added to stockmen's woes. Wrote one rural bard: "So we go to church on Sunday, And pray like hell for rain, But the Good Lord has gone walkabout, And things just stay the same."

Paul Chesley (all)

in tight, leaving at dusk to roam highways and towns for water and food. By day, herds of bony cattle graze roadside on public rights-of-way, their owners' pastures long ago exhausted, taken over by woody weeds.

Four years of El Niño had made the Wuruma Dam's reservoir a dusty canyonland. Hundreds of feet down, an iridescent green slick and thousands of motionless turtles and fish slowly stifling in the choking muck were all that remained of a water system upon which people for hundreds of square miles around had depended.

A sense of stillness, as though time had stopped, spread over the land of the Big Dry. For many bush dwellers there seemed little left to do but endure. "For now the banks are leaving people in houses with a roof over their head because no one'll buy their land," said fourth-generation cattleman Larry Acton, president of Queensland's United Graziers Association and past president of its Cattle Council.

"We Australians are now importing grain to feed ourselves and our animals for the first time ever," according to Ian Macfarlane, president of the Grains Council of Australia and the Queensland Graingrowers Association. "It adds to the demoralization that a country that normally exports 80 percent of its production is reduced to not feeding itself."

Already, the social fabric was unraveling. Farmers forced to shoot their own stock were turning guns on themselves, as suicide rates for rural areas showed a greater percentage increase than those of the cities. Families were wrenched apart, with wives seeking jobs in faraway towns and grown children opting out of the farming life.

"I have to advise young people to steer away from agriculture when I speak at colleges," says cattleman and farmer Sid Plant, a highly regarded community leader in Oakey.

"Ironically, rainfall can be one of the farmer's worst foes," says Queensland climatologist Roger Stone. "El Niño is a very cruel thing around here. It lets a little

Kerry Edwards/REUTERS/BETTMANN; Miguel Luis Fairbanks (opposite)

rain in now and then, here and there. It gives people false hope. They plant, stock up, build a new house, and that's the last rain they see for a year."

Sid Plant had resisted all of those impulses. Over the years he had paid close attention to the atmospheric pressure index of the Southern Oscillation, once buying cattle at bargain prices when falling air pressure suggested the El Niño would break. With the reverse now happening, it was a matter of survival. "Within two weeks I have to sell my cattle or move them to someone else's property." Plant had seen signs that El Niño was intensifying.

For eastern Australia's farmers and graziers, the next move would come down to an educated gamble. To hold or fold, bet thousands on a drought-resistant seed variety and a flukey shower or two? Like gambling, weather prediction is a game of odds, *(Continued on page 170)*

Nature Out of Kilter

Volunteers minister to Linketta, a young female koala, at the Koala Hospital and Study Centre in Port Macquarie, New South Wales (above). Koalas' preferred food—eucalyptus leaves—puts the animals in peril when drought and arson spark bush fires, as it did, disastrously, in 1994. Eucalyptus oils permeating the leaves exploded into flame, burning hundreds of koalas. Migrating herds of emus face another hazard during the El Niño-induced drought: a 900-mile fence protecting Western Australia farmland. Desperate for food and water, the flightless birds die as they smash into the fence (opposite).

Parched West Blazes

Firefighters mobilize for another 16-hour day on the fire lines of central Washington's Wenatchee National Forest and Chelan County, where 186,694 acres burned in 1994. Steep, rugged canyons funneled winds and flames across highways and up slopes, sometimes stymieing firefighters (below.) The Rat Creek fire scaled Icicle Ridge (above) after torching 12 houses near the town of Leavenworth.

Phil Schermeister (all)

Devastation, Renewal

Flames vault into the high canopy of a Douglas fir grove, gathering force for a treetop run (opposite). Fire suppression had resulted in a stand of closely spaced trees. Small trees form "fuel ladders," by which fire travels up to the crowns of taller trees. A dust devil whirls amid blackened remains (above); a month later, a Douglas maple shoot emerges from ash.

John Marshall (all)

Southern France's Black Tuesday

More tidal wave than torrent, the rain-swollen Ouvèze River crashes through the ancient town of Vaison-la-Romaine in southern France. After continual, violent thunderstorms, the river reared up into a 49-foot wall of water on what became known as Black Tuesday. It was September 22, 1992. In all, 38 people, including several campers in a nearby park, were swept to their deaths in the flash floods. A few days later, the upper-level trough responsible for the deluge in France brought flooding to Spain and Italy as well.

SYGMA

Some of the very same factors that set that tragedy in motion have occurred in very different settings—such as the American Great Plains. Like the Sahel, the Great Plains have a variable rainfall history, so variable, in fact, that westering explorers and settlers in the first half of the 19th century called the treeless expanses the Great American Desert.

Just as the Sahel's most marginal areas were settled when rains were abundant, successive waves of migrants spread across the American Plains during moist decades. Their sod-busting methods destroyed the soil's fertility and structure. Then, inevitably, came runs of dry years and a predictable exodus. The best-known occurred in the mid-1930s, when black tsunamis of dust thousands of feet high rolled across the sky and far beyond the Dust Bowl itself.

Midwestern farmers learned from their mistakes. They began planting windbreaks of trees and shrubs and plowing within the limits of the land's contours. More than anything, irrigation water from the giant underground Ogallala Aquifer insulated the region from severe droughts in the 1950s and 1970s.

During the 1970s, a world food crisis and its demand for grains brought into use land marginal for sustained grain production. Farmers destroyed their shelterbelts, and dust storms, visible from space, reared up again. Eventually, as parts of the Ogallala Aquifer continue to dry up and droughts recur periodically, Dust Bowl conditions are likely to recur from the sandhills of Nebraska to the high plains of Texas.

Clouds of grasshoppers, darkness at noon, fields flying away: The visual record, deeply etched in the American psyche, suggests biblical plagues. But those generally involve retribution for human actions, even as in the Hindu

Winter Floods Revisit Western Europe

Dampening holiday spirits, floods strike Western Europe two years in a row. By Christmas 1993, residents of Charleville-Mézières (opposite) are homeless for the holidays after relentless rains and snow-melting warmth trigger the "flood of the century." Some of the French city's people (below) hang on in January 1995, when the Meuse River submerges its old high-water mark. Floodwaters float groceries in Redon (above). A German headline groans, "Rain, Snow, Rain, Snow."

Eric Hadj/SIPA PRESS; Jobard/SIPA PRESS (upper); SIPA PRESS (opposite)

tradition, sins could spoil one's monsoon crops, and those of neighbors, too. Now, human beings play an ever-increasing role in shaping extremes of weather and climate. Dust Bowl farmers didn't call forth the drought, but they did create its dust storms. Likewise, people have not created the deluges that flood Mark Twain's "lawless stream," but levees have sped the Mississippi's straitened flow, backed up tributaries when the rains have come, and finally failed.

When it comes to weather, the effects of some of our efforts to bend the natural world to our will might be dubbed nature on the rebound. For example, when millions of acres of range and forestland in the Pacific Northwest and California went up in flame in the summer of 1994, a long stretch of bone-dry weather and high temperatures contributed to the area's readiness to burn, but years of fire suppression by people fanned the flames.

It seems that the balance of power in our restless relationship with climate may have shifted to the point where we can create our own. In fact, we have even invented new terms for new weather: "acid rain" and "heat islands," for example. Some scientists contend that a long-term, worldwide climate change popularly known as "global warming" is under way: Along with sea-surface temperatures, air temperatures around the globe have risen—by about 1°F in the last century. Sea levels have slowly been on the rise, as waters warm and glaciers melt.

Was this warming brought on by the accumulation of carbon dioxide and other heat-trapping gases in the atmosphere, called collectively greenhouse gases? We know that the level of carbon dioxide rose 20 percent between the 1880s and the 1980s, as industrial processes accelerated and people burned more fossil fuels—coal, oil, and gas—and cut down the tropical rain forests. We also know that the 1980s, with record-breaking heat, were distinctly warmer than any previous decade since record keeping began in the 1880s.

Arcen & Velden/ANP PHOTO

Scientists generally agree that if carbon dioxide keeps building up and temperatures continue to rise at their current rates, the world will be a very different place by the middle of the coming century. Weather and climate are likely to become more extreme, more variable, and therefore less predictable. Events like the catastrophic European floods of the fall and winter of 1994 and 1995, triggered in part by unseasonable warmth, may become commonplace. Along with more floods in some places, there may be more droughts in others. Storms may become more intense. Today's sweltering summers could seem springlike.

It won't take much—a few degrees Fahrenheit. Add to this the vast web of processes that link weather anomalies around the globe, and then figure in the weather's potential for sudden chaotic turns. Present-day computer models cannot generate every possible variation of new global climates. But they hint that the extreme weather shifts of the 1980s and 1990s could serve as a compelling preview.

Long Line of Defense

Dutch villagers heft sandbags along a Meuse River dike in January 1995. Some 250,000 people evacuated their homes during the worst flooding since Zeeland's North Sea dikes burst in 1953, killing more than 1,800. That event brought reinforcement of sea dikes. But in 1995, with the rain-swollen Rhine, Meuse, Waal, and other waterways overflowing, weak, waterlogged river dikes threatened to crumble and bring disaster to densely populated polders. Blame for Europe's flooding fell to more than heavy rains and melting snow. Urbanization and deforestation prevent the land from sponging water, adding to the river's volume. Germany alone loses 175 acres a day to asphalt and concrete. With similar forces at work worldwide, humans' role in weather disasters will grow in the future.

Coping with

Catastrophe

by Tom Melham

Rare blizzard, the worst in four decades, coated Jerusalem in 1992 with 16 inches of snow.

All Fall Down

Choices—and luck—affect homeowners everywhere. Choose wrong—or hit some bad luck—and the results can be catastrophic. For decades the home of actor Charles Laughton stood safely atop the Pacific bluffs in Malibu. Then, in January 1994, a mudslide yanked terra firma out from under it. Experts today recommend that home builders in such areas have both a geologist and a structural engineer check a site before construction begins. For the owner of a cottage at Kitty Hawk, North Carolina (below), who chose to build near the sea, luck ran out when Hurricane Gordon caused beach erosion in 1994.

It was my first time in Tokyo. The morning after an obligatory nocturnal prowl through the Ginza found me still in bed, half asleep, only dimly aware that, yes, the sun had risen. Then I felt the bed lurch slightly, almost vibrate, the motion irregular but as smooth and subtle as a well-executed lambada. Seconds passed before I realized that this was not some idiosyncrasy of the hotel's bedsprings; the room—the entire building—was moving around me.

Eyes snapped open. I immediately did the wrong thing: Rather than seek shelter under a table or desk, I went straight to the window and stared out. Other high-rises, like mine, were visibly swaying despite multiton loads, their windows filled with concerned faces also looking out. A few stared at me, most at the streets below. There was no apparent damage, no rubble of collapsed buildings, not so much as a broken pane of

Drew C. Wilson/*VIRGINIAN-PILOT*; Lynn Forman (opposite); AP/WIDE WORLD PHOTOS (preceding pages)

Blown Away

Nature also attacks her own: Like jackstraws, fallen trees litter the ground of the Francis Marion National Forest in South Carolina, following the ruinous passage of Hurricane Hugo in 1989. The storm damaged twice as much timber as the eruption of Mount St. Helens: some 6.7 billion board feet of sawlogs and 20 million cords of pulpwood. Unfortunately for the endangered red-cockaded woodpecker, 95 percent of the trees it nests in were damaged or destroyed.

glass or a fallen brick. Just that odd, surreal jiggling that began, lingered, and then ceased, all in less than ten seconds.

Yes, it was an earthquake. A light one, packing only a tiny fraction of the power of those magnitude-9-and-over monsters that ripped through Chile in 1960 and southern Alaska in 1964. All the same, its motion had been real. I was spellbound.

There is nothing quite like a little seismic activity to remind us that, despite millions of years of evolution and all our billions of technological achievements, nature still is the "baddest guy" around, with forces that continue to elude our control. A flooding Mississippi laughs at manmade levees, cyclones regularly batter Bangladesh, hurricanes hit the Atlantic

coast of the U.S., landslides and wildfires terrorize California, and volcanoes and earthquakes rule the Ring of Fire. What do we do about all this?

We've learned to track storms and monitor seismic shifts around the globe; we've also sorted out part of nature's logic behind El Niño and Tornado Alley. Satellites today not only look for changing cloud patterns but also measure changes in sea-surface temperatures and the extent of sea ice. The shuttle uses a radar imaging system to search for signs of land deformation that could presage geologic activity. We know that certain places are riskier than others—and we know why. We can forecast tomorrow's weather with considerable accuracy; we can even anticipate some volcanic eruptions in time to evacuate nearby communities—as occurred in the Philippines in 1991, just before Mount Pinatubo blew.

And science continues to penetrate the mysteries of Mother Nature. For example, Japanese researchers found, after the devastating earthquake in Kobe in 1995, that changes in the chemical composition of groundwater there correlated directly with the approach of the disaster. Such possible earthquake predictors will be tested in other quake-prone areas to help predict temblors soon enough to increase residents' readiness for them.

We've come a long way. But when it comes to mitigating many of nature's ragings, human actions consist largely of

Jay Browne/SOUTH CAROLINA FORESTRY COMMISSION

reaction: search and rescue, distributing food and other emergency supplies, cleaning up, rebuilding—often on the same disaster-prone sites, often without making new structures any better adapted than the old. Too often we respond to natural hazards much as humankind always has: not with foresight but with knee jerks.

Flooding, for example, is one of nature's more predictable hazards: We monitor both rainfall and river levels, at many sites. Yet in the U.S. floods continue to rank among the most deadly and destructive of all natural events. Why? Because we fail to recognize the potential hazards of high water, and often we are loath to leave floodplains to the floods.

Future Materials

When quakes strike rigid, brittle structures, rubble results. Here workmen raze a quake-damaged California freeway. Engineers fear many of the unreinforced concrete columns that hold up overpasses may not withstand future quakes. At the Kajima Construction Company in Tokyo (left), a worker tests a reinforced rubber cushion that, as part of a foundation, would flex back, forth, up, and down.

On the nightly news, images of disaster: A house turned to matchsticks by a tornado, the ghostly pall of a Pinatubo or a Mount St. Helens, roads erased by mudslides, a stretch of elevated freeway pancaked by an earthquake while adjacent sections are left seemingly intact.

We call such events natural disasters, but of course it's their human side that

Lynn Forman; Paul Chesley (opposite)

really interests us. Who died? How many people were injured? How many homes were lost? We crave to share in the human suffering and triumph, to meet and talk to survivors, to ask what it was like. We feel for the Bangladeshi peasant torn by the contrary emotions of relief and grief: He has just survived a devastating cyclone—but his wife and entire family were swept into the sea before his eyes. We marvel at the faith and gutsiness of a Romanian woman who, buried alive in her earthquake-shattered apartment building, endured nearly eight days without food or water or human comfort—and never gave up hope. We love heroes, and we know we will find them wherever natural disasters strike. The resolute couple, for example, standing beside the remains of what was their beachfront home, stubbornly proclaiming through their tears that, yes, they will rebuild—right here, bigger and better than ever.

Nature rages, humans respond. At times, our fear of earthquakes sparks improved building codes that promise to make new construction more quake-resistant. But, even then, often we fail to require improvement of existing structures. Similarly, each new flood elicits volunteer armies of sandbaggers—but never a total overhaul of our seriously flawed flood-control programs, which actually have increased flood danger rather than lessened it. Another hurricane warning prompts Floridians to board up and evacuate, as they have so often before. Wildfires in southern California bring homeowners to their rooftops with garden hoses. Not high-tech solutions, certainly not the best we can do.

But then our memories can be incredibly short. Thousands displaced when the Mississippi River overflowed its banks in 1993 were offered governmental help to relocate on higher ground. Some chose instead to stay on the floodplain—only to suffer a second disastrous washout just two years later. A similar, don't-look-back logic prevails at numerous other risky sites around the world: the thickly populated flanks of Italy's sleeping but far-from-dead Vesuvius, for example, or oceanfront developments on the American side of the Atlantic. Why erect inflexible houses on the moving sands of barrier islands? Why

On Tokyo Bay

Asking for trouble, workmen bulldoze landfill into Tokyo Bay. Such new ground, if not properly compacted, can turn to mush during earthquakes. Its proximity to the bay assures a high water table; quake vibrations compact the soft soils, causing the water to flow upward. A process called liquefaction occurs, turning the soil into quicksand and making it incapable of supporting buildings. Structures in such areas are vulnerable to collapse. They may tilt and sink into the liquefied soil. Storage tanks (opposite) on landfill crowd the edge of Tokyo Bay. Their presence concerns those aware of potential liquefaction.

locate in high-hurricane-risk areas—or on established seismic faults?

Another example of our failure to deal effectively with natural hazards: San Francisco's 1906 earthquake and fire totally obliterated nearly 500 city blocks; citizens there had never known such a terror. Yet they rebuilt as poorly, seismically speaking, as they had built before. Structurally weak architectural overhangs and parapets—proven dangers to passersby during quakes—were included again in the design of new buildings. Masonry was erected without reinforcing steel. Wood frames were not bolted to foundations. During a quake such houses simply slip off to one side when the ground starts shaking. Wooden row houses were built next to each other with no barriers to stop the spread of fire.

Certainly the everyday pressures of expediency and economics played a major role. San Francisco city officials lowered building and fire codes to pre-1906 criteria. Rebuild quickly, they figured, and leave the future to the future.

In the Marina District, rubble from buildings that collapsed in the 1906 quake was used for landfill in 1915 and was not properly compacted. Seventy-four years later, during the Loma Prieta quake, Marina row houses predictably slumped and buckled as the fill experienced what geologists call soil liquefaction.

Today, of course, we like to consider ourselves more responsible. The lack of any real damage to Tokyo on the day that I experienced my first (and only) earthquake seemed to confirm the popular image of Japan as a leader in quake-resistant construction. Its building codes are more stringent than those of many other nations; its use of flexible steel for framing buildings, concrete reinforced with steel, and other technology

Paul Chesley (both)

Paul Chesley (both)

Tall and Steady

When strong winds blow or the earth quakes, the Osaka World Trade Center Building shifts its center of gravity, much like a human would. A weight (left) atop the building slides—in any direction—in response to computer commands to counterbalance the structure. Past techniques merely tried to make buildings strong enough to endure such forces.

all help lessen quake damage and human injury. Its engineers have devised springy, rubber-and-steel foundations and can retrofit some older buildings with rubber "shock absorbers." So do their counterparts in the U.S., where cutting edge designs include new, "smart buildings" equipped with massive, moveable weights. Computerized controls automatically shift those weights in response to seismic stresses, theoretically enabling giant office buildings to weather quakes.

But in fact, my Tokyo quake was too small to be a true test. Also, despite all Japan's construction innovations and earthquake savvy, not even this nation is totally prepared. The Kobe quake killed 5,500 people, dispelling any idea of Japan's seismic invincibility. It also focused attention on what may happen when the inevitable "big one" hits Tokyo. The last one—a 7.9 magnitude shocker in 1923—killed 143,000 people.

Contemporary concerns for Tokyo go far beyond its superstructures—whether they are gleaming, modern high-rises (yet to be tested by nature) or the smaller, older, traditional style buildings. Mostly of wood and exempt from newer codes, these predominate in many neighborhoods of the city.

There is the additional concern for how broken water lines would severely impede fire fighting capabilities and how breaks in telephone and electrical service would spawn incredible chaos, especially since business and society in general today rely so totally on computers. Experts speculate that a major quake could turn Tokyo into a disaster zone.

But the same could be said of Memphis, Tennessee, or of Jonesboro, Arkansas. Although they lie far from any known plate boundary, both are staring down the barrel of a huge seismic cannon: the fault zone that gave rise to the gigantic New Madrid earthquakes of

From High Above

Eyes in the sky give scientists a hint at what's happening. Seen from the space shuttle *Discovery* in 1994, the Rabaul volcano on New Britain Island spreads a billowing white eruption plume 60,000 feet high. The darker layer beneath it is heavier ash. Twin images of Mount Pinatubo in the Philippines (opposite), taken nearly six months apart in 1994 by imaging radar aboard the shuttle *Endeavour,* help track deadly mudflows and the increasing erosion in ash and pumice deposits from the 1991 eruption.

1811 and 1812, which temporarily reversed the Mississippi's flow. This inherent risk moved Jonesboro to initiate stricter building codes in 1989. The city later repealed the earthquake ordinances until a committee had completed its study. Jonesboro then readopted the more stringent codes in 1990.

New York City approved new, earthquake-aware building standards in 1995, even though they would increase construction costs by an estimated 2 to 5 percent. Elsewhere, quake-readiness techniques have been projected to hike costs as much as 10 percent. A justifiable increase, perhaps, since public safety improves, but also quite a significant one.

Like balance sheets, bottom lines, and

NASA (all)

market fluctuations, the ragings of nature are a management thing. It's not as if we don't know they're coming; we just don't know where or when.

But imagine what our home planet would have been like if there had been no earthquakes or volcanoes—no plate tectonics or hot swirling mantle—no raging storms or droughts or atmospheric disturbances. By ferrying to the surface enormous amounts of water vapor, carbon dioxide, nitrogen, and other gaseous materials that once were chemically bound deep within molten rock, volcanoes helped create not only earth's atmosphere but also its oceans.

To this day volcanoes play major roles in the renewal of the earth. Their additions of lava rejuvenate the land both structurally and chemically; volcanic soils are among the richest on earth, while many mineral ores have volcanic origins.

Earthquakes, too, are as vital as they are destructive, continually rearranging the topography and uplifting new mountains, new cliffs, new coasts. Storms like hurricanes and monsoons sustain global systems, just as eating and drinking sustain us. If earth were totally peaceful, quiescent both geologically and meteorologically, it would be more like our moon. We fear and rail against nature's ragings, but they are a necessary part of life as we know it.

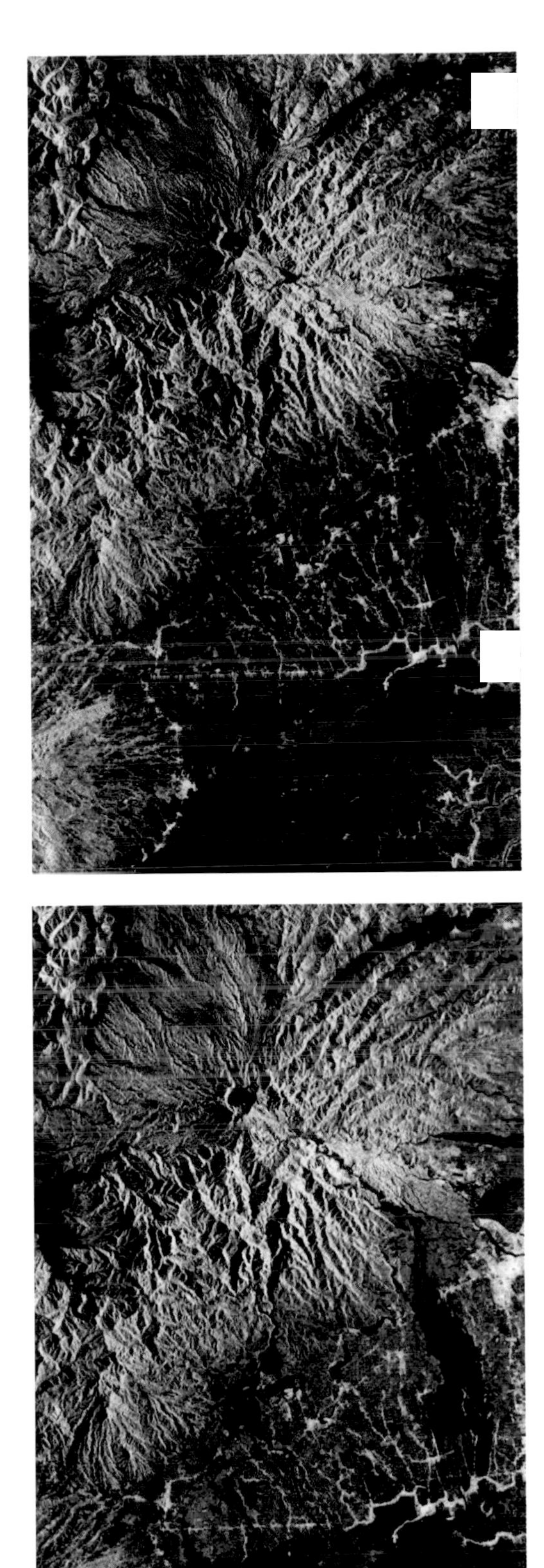

According to the United Nations International Ad Hoc Group of Experts, the current decade should witness about a million thunderstorms; maybe a hundred thousand floods; tens of thousands of landslides, earthquakes, wildfires, and tornadoes; and between several hundred and several thousand cyclones, droughts, tsunamis, and volcanic eruptions.

Such statistics—assuming they hold—make the '90s no stormier or quake-prone than other recent decades. The problem, however, is that the combined impacts of all these natural ragings are swelling ever larger. So it is that the United Nations is promoting the 1990s as the International Decade for Natural Disaster Reduction. The U.N.'s aim is to generate worldwide awareness of potential disasters and to stimulate efforts to minimize their effects. Its goal: a hazard-resilient world.

The U.N. program suffers from lack of funds, but it has already galvanized more than a hundred countries around the globe into making plans to study the potential problems of natural disasters and ways to prepare for them. Successful strategies often cited are Japan's national landslide prevention program, which drastically decreased landslide losses, and Bangladesh's satellite-based storm warning system that has reduced death tolls from cyclones considerably.

Much of the burgeoning worldwide population expansion is occurring in developing nations where demands for economic growth militate against practices that might reduce the human costs of natural disasters.

Economic costs, meanwhile, are escalating. In the U.S. natural disasters eat up 52 billion dollars yearly. The Kobe quake alone set Japan back 200 billion dollars. The world, as interdependent as communities and nations have become, approaches the point when outbursts of nature could trigger daunting economic disasters.

One of the worst scenarios: World markets could panic, setting off chain-reaction currency fluctuations and even failures—all because, back in the 1990s, we didn't do all we could to develop successful strategies to mitigate the effects of natural disasters. Until humans move beyond reacting to losses and, instead, lay plans to predict and minimize disasters, merely coping with catastrophe will remain basic to life on earth.

Roger Ressmeyer/CORBIS

On-Site Science in a Caldera

Otherworldly setting serves as a laboratory for an earth scientist deep in the steaming caldera of Mount Pinatubo. Hands-on inspection of the crater lake yields data unavailable from satellites. In such a setting, humans seem puny adversaries for the natural forces that periodically rage across the earth's face.

Notes on Contributors

As a staff writer, LESLIE ALLEN contributed to numerous Special Publications and authored *Liberty: The Statue and the American Dream.* Now freelance, she writes often on American social history and on environmental topics for the *New York Times, American Heritage,* and other periodicals and books.

THOMAS Y. CANBY, formerly a senior assistant editor for science on the staff of NATIONAL GEOGRAPHIC magazine, spent 31 years explaining scientific subjects to millions of readers. His work won numerous honors, including the AAAS–Westinghouse award.

In more than 30 years at the Geographic RON FISHER has been around the globe, covering subjects as diverse as volcanic eruptions in Colombia and the birds of Papua New Guinea. Fisher authored *The Earth Pack...Nature's Forces in Three Dimensions.*

Freelance author NOEL GROVE, a member of the Society's staff for 25 years, authored 28 articles for the NATIONAL GEOGRAPHIC and wrote the Special Publication *Wild Lands for Wildlife: America's National Refuges.* During his years of travel, he watched more than a dozen volcanoes erupting.

Senior writer TOM MELHAM has written extensively on environmental and outdoor adventure themes since joining the Society in 1971. Along the way he has chased cyclones in Bangladesh, earthquakes in Romania and Japan, and forest fires in Idaho.

Sam Mircovich/ REUTERS/ BETTMANN

A rough road gets rougher. Winter rains in 1994 brought mud sliding onto the Pacific Coast Highway in Malibu, California. The mud entered this couple's home (left), ran out through the back, then into the ocean.

Acknowledgments

The Book Division wishes to thank the many individuals, groups, and organizations mentioned or quoted in this publication for their help and guidance. In addition we are grateful for the assistance of the National Geographic Society News Collection and of the following individuals: Akira Hanajima, Steven Brantley, James P. Bruce, Gary Carver, Nicholas K. Coch, Emilie Cooper, Charlie A. Crisp, Dennis Decker, Andrea Donnellan, Daniel Dzurisin, Susanna Falsaperla, Sturla Fridriksson, Ken Gillies, Minard L. Hall, Arnold Hartigan, Walter W. Hays, Kazuhiro Ishihara, Kazuya Ohta, Jeff Keeler, Kevin E. Kelleher, George Kiladis, Koji Ishii, Dane Konop, Joe Leonard, Jr., Jim Luhr, Clay Manner, DeWayne Mitchell, Vincent Moreno, Tom Mullins, Don Parker, Kenneth Phillips, Ben Picou, Frank Pratte, Edward N. Rappaport, Betty Seibel, Kerry Sieh, Paul D. Smith, Dean Snow, Sister Myra Jean Sweigard, Edwin S. Taylor, Karen Terrill, Jim Verti, Brian Wadley, and Andrew Whitaker.

Additional Reading

The reader may wish to consult the *National Geographic Index* for related articles and books. The following titles may also be of interest:
Robert W. Decker and Barbara G. Decker, *Mountains of Fire* and *Volcanoes;* Margaret Fuller, *Forest Fires;* Michael Glantz, ed., *Drought Follows the Plow;* Thomas P. Grazulis, *Significant Tornadoes 1680-1991;* Gladys Hansen and Emmet Condon, *Denial of Disaster;* Richard A. Keen, *Skywatch East;* David Roland, ed., *Nature on the Rampage;* Keith Smith, *Environmental Hazards—Assessing Risk and Reducing Disasters;* Kaari Ward, ed., *Great Disasters;* Robert Wenkam, *The Edge of Fire;* Jack Williams, *The Weather Book;* and Thomas L. Wright and Thomas C. Pierson, *Living With Volcanoes.*

Library of Congress CIP data
Raging forces : earth in upheaval / prepared by the Book Division, National Geographic Society.
p. cm.
Includes index.
ISBN 0-7922-2736-0 (reg.). - ISBN 0-7922-2965-7 (deluxe)
1. Natural disasters. I. National Geographic Society (U.S.). Book Division.
GB5014.R35 1995
363.3'4-dc20 95-36295
CIP

Composition for this book by the National Geographic Society Book Division with the assistance of the Typographic section of National Geographic Production Services, Pre-Press Division. Printed and bound by R. R. Donnelley & Sons, Willard, OH. Color separations by Digital Color Image, Cherry Hill, NJ; Graphic Art Service, Inc., Nashville, TN; Lanman Progressive Co., Washington, DC; and Penn Colour Graphics, Inc., Huntingdon Valley, PA. Dust jacket printed by Miken Systems, Inc., Cheektowaga, NY.

Preparing for Disaster

DISASTERS can happen anywhere: Some strike quickly and with little warning. Basic services such as electricity, water, gas, and telephones may be out for days. Plan ahead. **Disaster preparedness works.** Here are some things you can do to prepare for a sudden emergency.

FINDING OUT WHAT COULD HAPPEN

1. Learn about the kinds of natural disasters most likely to happen in your community by contacting the local emergency management office or American Red Cross. Ask how to plan for each kind of disaster.

2. Find out your community's warning signals and the evacuation routes.

3. Ask about the emergency response plans for your child's school or day-care center. Talk about disaster plans at your workplace.

CREATING A DISASTER PLAN

1. Talk with family members about the various types of dangers and the way to respond to each. Make sure everyone knows what to do and where to go in an emergency. Share responsibilities and work together as a team.

2. Involve children in these family-protection meetings. Explain to your family the hazards of fires, earthquakes, or tornadoes; and teach kids basic safety rules. Children with knowledge and assigned tasks usually feel less vulnerable. Tell your children about steps to take to reduce effects of disasters.

3. Determine the safe spots in your home to ride out danger. Draw a floor plan and mark two escape routes from each room. Practice what you discuss.

4. Post emergency phone numbers and teach children how and when to call 911. Agree to turn on a radio to receive official information in an emergency.

9. Show family members how to turn off gas, water, and electricity at main controls and switches.

10. Plan how family members would stay in touch in case of separation in a disaster. Select two meeting places. One should be near your home in case of sudden emergency. The second should be outside your neighborhood in case you can't return home.

11. Pick a friend or relative outside the area for the family-contact person. Often it's easier after a disaster to call long distance. All family members should call this person and tell them where they are. Everyone must know the contact's telephone number.

12. Make arrangements for your pets. Health regulations do not permit pets in public shelters, so you will need to leave them at home with plenty of food and water. Call the Humane Society for other options.

ASSEMBLING A DISASTER-SUPPLIES KIT

Once disaster strikes, you won't have time to shop or search for supplies. Put together a disaster supplies kit. If you must evacuate quickly, you will have what you need prepared and ready to take with you. Here are some important items to include:

1. A waterproof container for your supplies—a camping backpack, a covered trash container, or a duffel bag.
2. A battery-powered radio, a flashlight, and plenty of extra batteries.
3. A supply of water for each person. Store it in clean, tightly sealed plastic soft drink bottles.
4. Foods that won't spoil, are light in weight, and require no refrigeration. Select nonperishable packaged or canned food and include a manual can opener.
5. A first aid kit, along with nonprescription and prescription medications that family members need.
6. A change of clothing for each person. Include rain gear, sturdy shoes or boots, hats, and gloves.
7. Blankets or sleeping bags.
8. Items required by infants, the elderly, or the disabled.

Store the kit in a place known to all family members. Every six months, change the water supply and check on the food. Replace batteries and keep all supplies up to date.

RETURNING HOME AFTER A DISASTER

Be very cautious when you re-enter your home. It might have suffered structural or water damage.

1. Wear sturdy shoes or boots and gloves to protect yourself from broken glass or other debris.

2. Use a battery-powered flashlight for light. Don't use a gas lantern, matches, or candles. Leaking gas or other flammable materials may be present.

3. Look for cracks in the roof, foundations, and chimneys. If it appears the building may collapse, leave immediately. Be alert to loose boards and slippery floors.

4. Do not turn on electrical switches if you suspect any damage.

5. Keep an eye open for the unexpected—poisonous snakes, for example. Use a stick to look through debris and turn things over.

6. Sniff for gas leaks. If you suspect a leak, turn off the main valve, open windows, and leave quickly. Call the gas company and report the problem.

7. Check for fires or fire hazards along with other household dangers.

8. Inspect the water and sewage systems. Shut off any damaged utilities.

9. Clean up spilled materials such as medicines, bleaches, gasoline, and other flammable liquids.

The first concern after a disaster is the health and safety of your family. Families who prepare can reduce the tension and anxiety and even losses by educating themselves ahead of time. Contact your local emergency management offices or Red Cross and obtain materials on ways to prepare for specific events such as hurricanes, fires, or winter storms.

These guidelines were excerpted from materials prepared for the public by the American Red Cross and the Federal Emergency Management Agency.

Index